65 Advances in Polymer Science

Fortschritte der Hochpolymeren-Forschung

Analysis/Networks/ Peptides

With Contribution by
K. Bode, I. Kuriyama, J. E. Mark
F. Maser, M. Mutter, Y. Nakase,
A. Odajima, V. N. R. Pillai, J. P. Queslel,
H. W. Siesler

With 143 Figures and 21 Tables

Springer-Verlag
Berlin Heidelberg GmbH 1984

ISBN 978-3-662-15281-2   ISBN 978-3-540-39029-9 (eBook)
DOI 10.1007/978-3-540-39029-9

Library of Congress Catalog Card Number 61-642

© Springer-Verlag Berlin Heidelberg 1984

Originally published by Springer-Verlag Berlin Heidelberg New York Tokyo in 1984
Softcover reprint of the hardcover 1st edition 1984

Typesetting: Th. Müntzer, GDR;

2154/3020–543210

Editors

Editorial

With the publication of Vol. 51, the editors and the publisher would like to take this opportunity to thank authors and readers for their collaboration and their efforts to meet the scientific requirements of this series. We appreciate our authors concern for the progress of Polymer Science and we also welcome the advice and critical comments of our readers.

With the publication of Vol. 51 we should also like to refer to editorial policy: *this series publishes invited, critical review articles of new developments in all areas of Polymer Science in English (authors may naturally also include works of their own)*. The responsible editor, that means the editor who has invited the article, discusses the scope of the review with the author on the basis of a tentative outline which the author is asked to provide. Author and editor are responsible for the scientific quality of the contribution; the editor's name appears at the end of it.

Manuscripts must be submitted, in content, language and form satisfactory, to Springer-Verlag. Figures and formulas should be reproducible. To meet readers' wishes, the publisher adds to each volume a "volume index" which approximately characterizes the content.

Editors and publisher make all efforts to publish the manuscripts as rapidly as possible, i.e., at the maximum, six months after the submission of an accepted paper. This means that contributions from diverse areas of Polymer Science must occasionally be united in one volume. In such cases a "volume index" cannot meet all expectations, but will nevertheless provide more information than a mere volume number.

From Vol. 51 on, each volume contains a subject index.

Editors   Publisher

Table of Contents

Rheo-Optical Fourier-Transform Infrared Spectroscopy: Vibrational Spectra and Mechanical Properties of Polymers

H. W. Siesler
Bayer AG, Werk Dormagen, Research & Development,
Postfach 1140, D-4047 Dormagen, FRG

The advantages of Fourier-Transform infrared (FTIR) spectroscopy over conventional dispersive instrumentation have revitalized the utilization of vibrational spectroscopy in polymer research. In the field of rheo-optics primarily the rapid-scanning capability of the FTIR technique has contributed to a more detailed insight into the technologically important process of polymer deformation. It is the purpose of this review to present the relevant instrumental and theoretical background of polymer deformation and relaxation studies by simultaneous FTIR spectroscopic and mechanical measurements and to summarize selected experimental results obtained by this technique.

Advances in Polymer Science 65
© Springer-Verlag Berlin Heidelberg 1984

1 Introduction

The mechanical properties of polymeric materials are of considerable importance for their engineering applications. In this respect the understanding of the molecular mechanisms involved in polymer deformation is a necessary prerequisite for a reasonable structure-property correlation.

Generally, from spectroscopic data such as frequency position, band shape, intensity and dichroism of specific absorption bands conclusions can be derived in terms of the applied mechanical stress and the state of order and orientation of the polymer under investigation. An extremely powerful method for the study of transient phenomena in polymer deformation and relaxation is rheo-optics which describes the relation between stress, strain and an optical quantity (for example birefringence, infrared absorption, light scattering, X-ray diffraction) measured simultaneously with stress and strain as a function of time [1-5]. In a given rheo-optical method therefore, a mechanical test is combined with one of these various types of optical measurements.

The advent of rapid-scanning FTIR systems has tremendously expanded the application of vibrational spectroscopy in the field of rheo-optics [6-11] and will certainly stimulate further progress in this research area. A similar development can presently be observed for X-ray diffraction since the availability of synchrotron radiation [12-15].

The theory and instrumentation of FTIR spectroscopy have been thoroughly treated in the literature [6, 16-18] and only some brief comments will be made here. Primarily two advantages have contributed to the break-through of the interferometer-based FTIR technique over conventional dispersive IR spectroscopy. The *multiplex* or *Fellgett's* advantage arises from the fact that in the FTIR spectrometer all frequencies of the polychromatic radiation contribute simultaneously to the interferogram whereas in a dispersive spectrometer the radiation is detected successively in small wavenumber increments. The *Jacquinot* or *throughput* advantage implies that the power of the source is more effectively utilized in the optical system of an interferometer with rotational symmetry about the radiation direction. Apart from the exploitation of the dedicated computer for the evaluation, representation and documentation of the spectral data these features have led to the superior performance of the FTIR technique with respect to increased signal-to-noise ratio, higher energy throughput and rapid scan (presently down to about 0.1 second for the entire mid-infrared region). Owing to the long scan duration of dispersive instruments the characterization of molecular changes in polymer deformation processes was so far restricted to stepwise elongation procedures or to the observation of narrow wavenumber regions in comparatively large time intervals during continuous elongation. Hence, the conclusions derived on a microstructural scale from spectroscopic investigations could only to a limited extent be satisfactorily correlated with the macroscopic properties evaluated from the stress-strain diagrams (e.g. elastic modulus, yield point, strain hardening, stress hysteresis, stress relaxation). This situation has been largely improved because spectroscopic parameters of the conformation, crystallization and orientation of the polymer chains can now be monitored by rapid-scanning FTIR systems on-line to the deformation process in very small strain intervals relative to the total elongation. Additionally, the influence of stress relaxation on particular absorption bands has become accessible to both the slow and fast decay region.

Further improvements in the time resolution of rheo-optical FTIR spectroscopic investigations to the millisecond-range have been quite recently reported by Hsu et al. [19,20] and Koenig and Fateley [18,21]. In their applications of time-resolved FTIR spectroscopy ordered interferometric sampling techniques [22,23] are incorporated to study the structural consequences of repetitive oscillatory strains imposed on polymer films. However, at present, not enough reliable data are available to assess the real potential of this technique.

2 Experimental and Software

The rheo-optical FTIR spectra presented in this review have been recorded on a Nicolet 7199 FTIR spectrometer equipped with a 1280 Nicolet 64 K computer.

The electromechanical apparatus constructed for the simultaneous measurement of FTIR spectra and stress-strain diagrams during elongation, recovery, and stress relaxation of polymer films is shown in Fig. 1. In such a stretching machine the polymer sample under examination can be uniaxially drawn at variable elongation rates while mounted in the sample compartment of the FTIR spectrometer. The specimen to be tested is held between two clamps which are movable by means of a spindle drive and which are attached to force and displacement transducers, respectively. Thus, two voltages proportional to the two mechanical quantities are directly recorded and digitally converted and taking into account the initial cross-section of the sample under investigation, the stress-strain diagram is obtained. For orientation measurements the polarization direction of the incident radiation can be alternately adjusted parallel and perpendicular to the stretching direction by a pneumatically rotatable polarizer unit which is also controlled by the computer. Specific values of the dichroic ratio (see below) for any absorption band at small strain intervals may then be obtained by relating the mean absorbance value of two subsequent parallel polarization spectra to the absorbance value of the corresponding perpendicular polarization spectrum and vice versa. In the spectrometer compartment the stretching machine is mounted on a x-y-stage for the variable alignment of the investigated sample area in the machine and transverse directions of the sample. The exact position of the illuminated sample area which is primarily important for the examination of polymers elongating by formation of a neck can be monitored with the built-in laser. The additional installation of a heating device in a closed system offers the possibility to study deformation and stress relaxation under controlled temperature conditions. To illustrate the technique the polarization spectra of the $\nu(C=O)$ and $\delta(NH) + \nu(CN)$ vibration region of a polyester urethane recorded during elongation in a loading-unloading cycle have been inserted in Fig. 1 along with the corresponding stress-strain diagram.

By exploiting the automated information processing capability of the dedicated computer in the FTIR system much of the routine analysis of spectra series produced in rheo-optical measurements has been alleviated. In this respect BASIC software was developed [24] for the evaluation and representation of the spectroscopic data in terms of various parameters:

a) Routine for the calculation of absorption intensities

The intensities of selected absorption bands (which, for example, are characteristic of a specific conformation or crystalline modification of the polymer) can be evaluated as peak maximum or integrated intensity between specified wavenumber

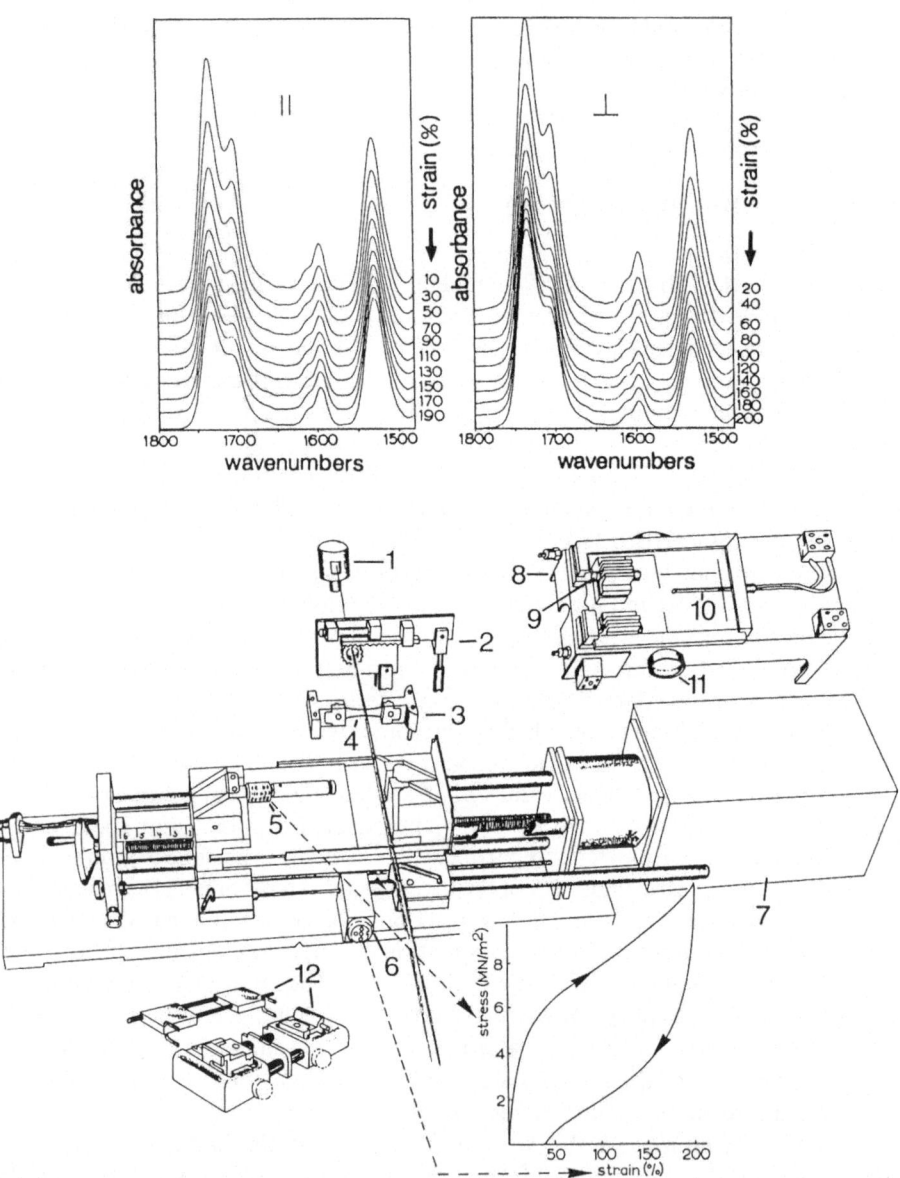

Fig. 1. Film stretching machine: (1) FTIR detector, (2) pneumatically rotatable polarizer unit, (3) clamp, (4) polymer film sample, (5) stress transducer, (6) displacement transducer, (7) driving motor, (8) heating accessory, (9) cartridge heater, (10) temperature control, (11) KBr window, (12) specimen preparation and transfer device

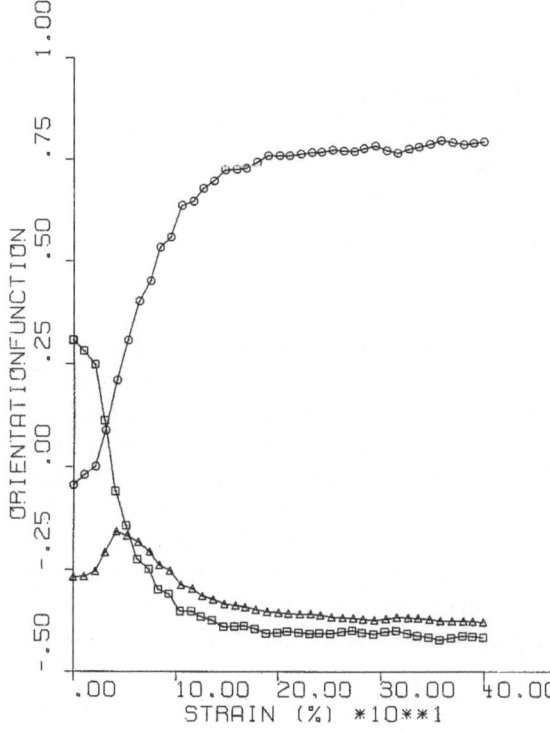

Fig. 2. Example of computer-plot of the crystal-axes orientation functions of orthorhombic polyethylene as a function of strain: $-\square-(f_a), -\triangle-(f_b), -\bigcirc- (f_c)$; (see also chapter 7.1.1)

limits and corrected for changes in sample thickness with an appropriately chosen reference band.

b) Routine for the calculation of dichroic ratios, orientation functions or structural absorbances (see below) in polarization measurements

In rheo-optical polarization measurements spectra are recorded alternately with light polarized parallel and perpendicular to the direction of elongation. Based on the peak maximum or integrated intensity this routine determines the dichroic ratio or orientation function (in the case of well-defined transition moment directions) of specified absorption bands for each spectrum by correlating the subsequently determined intensity values (Fig. 2). Alternatively, the structural absorbance which eliminates the influence of changing orientation on the actual intensity of a particular absorption band in rheo-optical experiments with large elongations by differently weighting the parallel and perpendicular polarization spectra can be evaluated. Here too, ratioing against a reference band automatically compensates for changes in sample thickness.

c) Peak search routine

The peak maxima of selected absorption bands can be automatically searched in a series of spectra with the aid of this program.

d) Plot routines

Upon data processing with the routines described under a–c for a series of spectra taken as a function of time, temperature, stress or strain the individual values can be plotted as a function of any of these variables in an operator-selected format (Fig. 2).

e) Spectra plot routine

Finally, this program [25] allows the stacked, overlap-free representation of a series of spectra in an operator-selected format (see, for example, Fig. 8).

3 Vibrational Spectroscopy of Stressed Polymers

When a tensile stress is applied to a polymer sample several mechanisms may contribute to changes in the vibrational spectrum of the polymer under examination. Thus, significant intensity variations can be observed as a consequence of phase transitions, strain-induced crystallization or orientation of the polymer chains. Apart from these effects it has been demonstrated in a series of investigations on the molecular mechanics of stressed polymers, both theoretically and experimentally, that the frequency and shape of absorption bands — predominantly those which contain contributions of skeletal vibrations — are stress sensitive [26-37]. This sensitivity of molecular vibrations to mechanical stress has been interpreted in terms of different mechanisms such as quasielastic deformation (reduction of force constants due to bond weakening under stress), elastic bond stretching or angle bending and conformational variations [26, 31, 32, 36].

Typically, the following spectroscopic changes have been observed upon application of stress:

1. Asymmetric distortion of the low-frequency wing of an absorption band
2. Wavenumber shifts $\Delta\bar{\nu}$ which were found almost exclusively into the direction of lower frequencies and which appeared to be a linear function of the applied stress σ [30, 32, 35, 36]:

$$\Delta\bar{\nu} = \alpha\sigma \tag{1}$$

where the stress sensitivity factor α has been shown to depend on the morphology of the investigated polymer [35, 36].

For the accentuation of these small differences in the spectra of the stressed and unstressed polymer the absorbance subtraction technique has proved particularly useful. In Fig. 3 this is illustrated with reference to the 972.5 cm^{-1} absorption band of the $\nu(O-CH_2)$ skeletal vibration of polyethylene terephthalate. Fig. 3a shows the shape of this absorption band for the unstressed and stressed (300 MN/m^2) polymer. In the difference spectrum (see Fig. 3b) the shift of the peak maximum toward lower wavenumbers and the low-frequency tailing are reflected by a pronounced asymmetrical dispersion-shaped profile.

The band distortion has been interpreted in terms of a non-uniform distribution of the external load in the regions of different state of order resulting in a non-symmetrical displacement of the individual atomic absorption frequencies about the maximum of the band [30, 31, 35-37]. Thus, Vettegren [36] postulated that over-stressed segments are predominantly located in the amorphous regions of the polymer.

In a recent publication Wool [37] has derived that for *Lorentzian* absorption bands

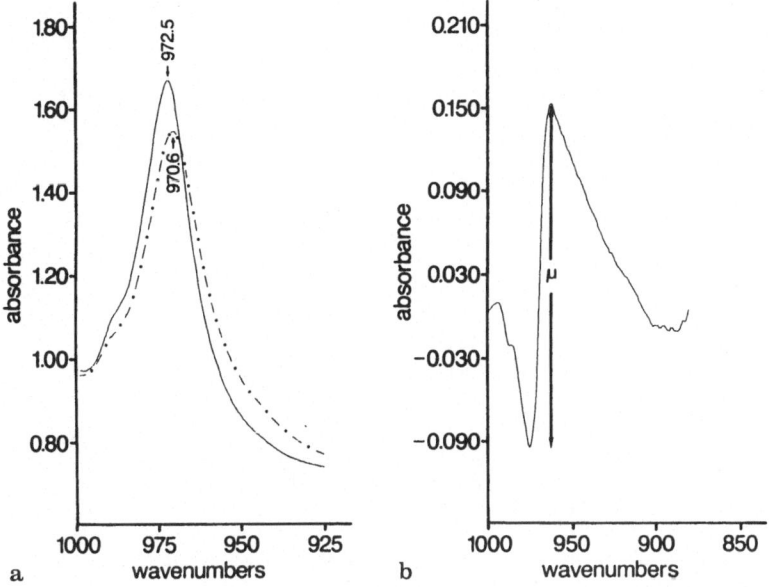

Fig. 3a and b. Effect of mechanical stress on the 972.5 cm^{-1} ν(O–CH$_2$) absorption band of polyethylene terephthalate: **a** ——— unstressed polymer, —·—·—·— stressed (300 MNm^{-2}) polymer; **b** difference spectrum: stressed polymer — unstressed polymer and corresponding peak-to-peak intensity (μ) (see text)

the relationship between the peak-to-peak intensity μ in the difference spectrum (see Fig. 3b) and the small wavenumber shift $\Delta\bar{\nu}$ is well approximated by:

$$\mu \simeq 1.3 \frac{A}{\Gamma} \Delta\bar{\nu} \tag{2}$$

where A is the absorbance and Γ is the half-band width.

The majority of results so far available in this field have been derived from spectra recorded on dispersive instruments. In order to test the validity of Eq. (1) several independent experiments have been performed in which a highly oriented polyethylene terephthalate film is subjected stepwise to increasingly higher stresses in the stretching machine illustrated in Fig. 1 and FTIR spectra are taken at the relaxed stress levels with unpolarized radiation. In Fig. 4 the wavenumber shifts of the ν(O—CH$_2$) absorption band are plotted as a function of the applied stress for different experiments. From these systematic investigations, however, no linear relationship between the wavenumber shift $\Delta\bar{\nu}$ and the applied stress σ could be derived whereas a fairly good index of determination (0.98) was obtained for the power function:

$$\Delta\bar{\nu} = \alpha \cdot \sigma^k \tag{3}$$

with $\alpha = 0.034$ and $k = 0.722$, where k represents the deviation from the previously postulated linearity. In Fig. 5 the difference spectra (stressed polymer — unstressed

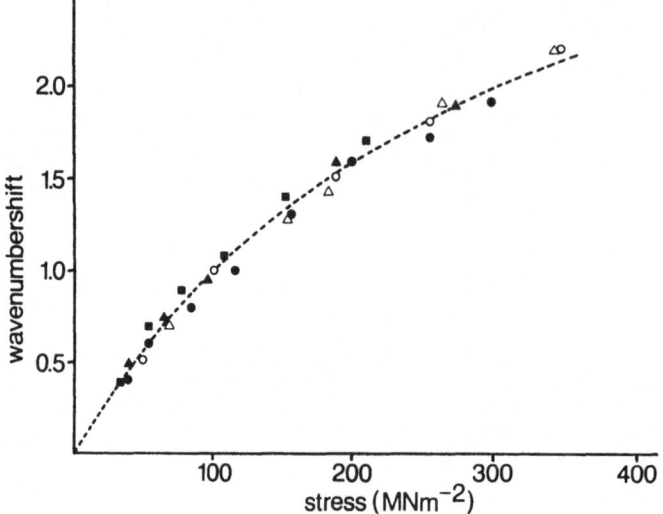

Fig. 4. Wavenumber shift $\Delta\bar{v}$ of the $v(O-CH_2)$ absorption band of polyethylene terephthalate as a function of the applied stress for independent experiments

polymer) are shown for a series of stress levels in the wavenumber interval 1000 to 890 cm^{-1}. Reasonably good agreement with Eq. (2) was established for the relation between the peak-to-peak intensity μ and the wavenumber shift $\Delta\bar{v}$.

Generally, these observations suggest the potential of selected absorption bands as a probe to monitor mechanical stresses and stress-induced molecular phenomena in polymers. In this respect particularly FTIR spectrometers with their high wavenumber accuracy may contribute to a more reliable experimental basis for the evaluation of such effects. Nevertheless, it has to be kept in mind, that any relationship

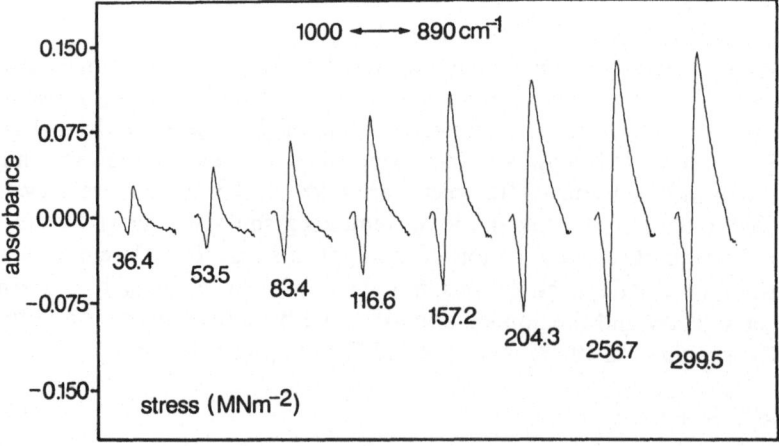

Fig. 5. Difference spectra (stressed polymer — unstressed polymer) for selected stress levels in the 1000—890 cm^{-1} wavenumber region (see text)

derived from the experimental results strongly depends upon the thermal and mechanical history of the polymer sample and cannot be transferred to other systems without further examination.

Polymeric engineering materials subjected to repeated cyclic or random mechanical stresses exhibit a deterioration of material properties termed fatigue [38]. Here too, minute changes in the FTIR spectra caused by the mechanical treatment can be enhanced by difference spectroscopy of the fatigued and unfatigued specimen. Through simultaneous application of electron microscopy, X-ray diffraction and FTIR spectroscopy Sikka [39-41] has studied the molecular response of amorphous polymers under tensile stress, cyclic fatigue and crazing conditions. Thus, for polystyrene small changes in the infrared bands assigned to various vibrational modes of the phenyl side groups have been detected as a consequence of the fatigue process and have been interpreted in terms of modified intra- and intermolecular interactions of the polymer chains. Fatigue-induced phenomena upon stress-cycling at different temperatures have also been reported for polyimide [42, 43].

Several electron spin resonance (ESR) investigations have established that free radicals are formed as a result of interatomic bond scission under mechanical stress and during fracture of polymers [38, 44-49]. The application of ESR spectroscopy, however, is limited by the reduction of radical population by abstraction, combination, and termination reactions especially in the presence of air or at elevated temperatures. IR spectroscopy has proved a valuable alternative in detecting the new end groups resulting from chain rupture by difference spectroscopy of the stressed or fractured sample against the unstressed specimen. Thus, Zhurkov [44] and Tabb [50] have estimated the concentration of ruptured bonds in mechanically loaded or fractured polyethylene and polypropylene by evaluating the intensity of characteristic absorption bands. It was shown that under tension and after fracture the absorption bands at 909, 965, 1378, and 1735 cm^{-1} increased in intensity. These absorption bands can be assigned, respectively, to $\gamma_w(CH_2)$ vinyl, $\gamma_w(CH)$ vinylene, $\delta_s(CH_3)$ methyl, and $\nu(C=O)$ aldehyde vibrations. The relation between viscometry, ESR and IR estimates of bond rupture and their relevance to mechanical properties of several polymers have been considered by Crist [51]. Quite recently the number of chain scissions resulting from mechanical degradation of polyethylene has been determined from FTIR analysis of new end group concentrations by Fanconi et al. [52]. Interestingly, these authors find that the concentration of new end groups in fractured polyethylene (about $5 \cdot 10^{17}/cm^3$) is lower by an order of magnitude than the values previously reported [44, 50] and that the number of chain scissions per free radical (~ 100) is lower by one to two orders of magnitude.

4 Orientational Measurements in Polymers Using Infrared Dichroism

Orientation in polymeric materials is a phenomenon of great practical and theoretical interest. The measurement of orientation provides a basis for the description of both the process whereby the oriented polymer is formed as well as the accompanying modification of its physical properties.

Among other techniques such as X-ray diffraction, birefringence, NMR spectroscopy, polarized fluorescence, Raman depolarization and sonic techniques the measurement of infrared dichroism with polarized radiation is one of the most frequently applied tools for the analysis of anisotropy in polymeric materials. Detailed accounts on the theory of infrared dichroism are available in the literature [6, 53–56] and therefore only the basic principles will be briefly outlined here.

When an oriented polymer is investigated with linearly polarized radiation the absorbance a of a single group in the polymer chain is proportional to the square of the scalar product of its transition moment vector M and the electric vector E of the incident polarized radiation:

$$a \propto (ME)^2 = (ME)^2 \cos^2 \gamma \tag{4}$$

where $|M| = M$, $|E| = E$, and γ is the angle between the transition moment and the electric vector. Thus, maximum absorption takes place when the electric vector is parallel to the transition moment of the vibrating group, but no light will be absorbed when its electric vector is perpendicular to the transition moment. The actually observed absorbance A is equal to the sum of absorbance contributions of all structural units:

$$A \propto \int_n (ME)^2 \, dn \tag{5}$$

where the integral refers to the summation of all molecules. When these atomic groups and their associated transition moments are randomly oriented within the polymer, the measured absorbance is independent of the polarization direction of the incoming light. For anisotropic distribution of the tranition moments, the observed net absorbance varies with the direction of the electric vector E of the polarized radiation. The effect of anisotropy on a particular absorption band in the IR spectrum of a polymer is characterized by the dichroic ratio R

$$R = \frac{A_\parallel}{A_\perp} \tag{6}$$

where $A_\parallel$ and $A_\perp$ are the integrated absorbances measured with radiation polarized parallel and perpendicular to the draw direction, respectively. Under the assumption of equivalent band shapes the integrated absorbances may be replaced by the absorbances measured at the peak maxima.

If, on the other hand, the intensity of an absorption band exclusive of contributions due to the orientation of a highly elongated polymer is required for quantitative analysis, the so-called structural absorbance A_0 has to be evaluated from polarization measurements:

$$A_0 = \frac{(A_u + A_v + A_w)}{3} \tag{7}$$

where A_u, A_v and A_w are the absorbance components in the transverse, thickness and stretching direction, respectively, of the sample geometry. For uniaxial orientation A_0 reads:

$$A_0 = \frac{(A_{\parallel} + 2A_{\perp})}{3} \tag{8}$$

The determination of the dichroic ratio is of significant importance for three aspects:
1. Assignment of absorption bands
2. Determination of molecular geometry
3. Characterization of the relative polymer chain orientation

It must be emphasized, however, that orientational measurements by the IR technique require a knowledge of the respective transition moment direction relative to the chain axis. Owing to the small mass of the hydrogen atom, for example, the transition moment directions of certain stretching vibrations such as C—H or N—H will be localized along the respective chemical bond. The participation of the hydrogen atom in a hydrogen bond, however, may result in distortion of the transition moment direction. Because of the large force constants involved, transition moments of stretching modes belonging to multiple bonded atoms will also closely coincide with the bond direction [e.g. $\nu(C=O)$, $\nu(C\equiv N)$] as long as no significant delocalization of the bonding electrons occurs. Strong mechanical and electrical coupling of the C=O and C—N stretching vibrations in the planar amide group —CONH— cause appreciable deviation (approximately $20°$) of the $\nu(C=O)$ transition moment from the corresponding bond direction [56, 57].

During elongation of a semicrystalline polymer several different processes, such as elastic deformation of the original spherulitic superstructure, transformation of a spherulitic into a fibrillar structure, plastic deformation of microfibrils by slippage processes and elongation of molecular chains in the amorphous regions may occur simultaneously, successively or partly superimposed [4, 58–62]. One of the advantages of IR spectroscopy in this respect is that under certain conditions these complex processes may be resolved into the individual components. Thus, in general, the crystalline and amorphous phase of a semicrystalline polymer can be characterized separately by evaluating the dichroic effects of absorption bands which are peculiar to these regions. Furthermore, vibrational transition moments of symmetrical molecules are parallel or perpendicular to their symmetry elements, and therefore dichroic measurements on absorption bands of the crystalline phase can be employed to study the orientation of certain crystallographic axes of the conformationally regular structure within the crystalline regions. When more than one polymer chain runs through the unit cell, the splitting of conformational regularity bands in two components which may be polarized along different crystal axes can be observed as a consequence of intermolecular forces.

An alternative technique for the separation of the crystalline and amorphous contributions to the absorption bands of functional groups containing labile hydrogen atoms (e.g. —OH, —NH) is deuterium exchange (see below). The principal advantage of this experimental method is that it offers the possibility of studying independently the preferential alignment of chemically identical structural units in different phases of the polymer.

For the practical analysis of orientation phenomena in polymers mathematical relations between the parameters of distribution function models of the chain axes and the dichroic ratio have been derived and can be evaluated in terms of the experimentally measured values. Although this approach is restricted to relatively simple models, it has been extensively applied to establish the basic concepts of orientation mechanisms in a great variety of polymeric systems. In what follows the results of such calculations for some selected distribution models will be discussed.

For the simplest model of perfect uniaxial order it is assumed that the polymer chains are all oriented parallel to the draw direction, and that the transition moments associated with the vibrations of the absorbing groups lie in a cone with a semiangle ψ and the draw direction as axis. The dichroic ratio may then be expressed by [57]:

$$R_0 = 2 \cot^2 \psi \tag{9}$$

As ψ varies from 0 to $\pi/2$, R_0 varies from ∞ to 0, and no dichroism ($R_0 = 1$) will be observed for $\psi = 54°44'$.

In practice the orientation of the molecular chains is never perfect and the real situation can be described by introducing a factor f, defined by pretending a fraction f of the polymer to be perfectly uniaxially oriented, while the remaining fraction $(1 - f)$ is randomly distributed. The dichroic ratio R is then given by [63]

$$R \qquad \frac{f \cos^2 \psi + (1/3)(1 - f)}{(1/2)f \sin^2 \psi + (1/3)(1 - f)} \tag{10}$$

where ψ is the angle between the transition moment of the absorbing group and the chain axis of the polymer molecule. Equation (10) may be put into a different form:

$$f = \frac{(R - 1)(R_0 + 2)}{(R_0 - 1)(R + 2)} \tag{11}$$

Here, R is the measured dichroic ratio of the investigated absorption band and R_0 is defined as in Eq. (9). For an absorption band having its transition moment parallel ($\psi = 0°$) or perpendicular ($\psi = 90°$) to the chain axis f then reads:

$$f_{\parallel} = \frac{(R - 1)}{(R + 2)} \tag{11a}$$

and

$$f_{\perp} = -2\frac{(R - 1)}{(R + 2)} \tag{11b}$$

As an alternative to Eq. (10) the effect of imperfect orientation may be represented by supposing all the molecular chains to be displaced by the same angle θ from parallelism with the draw direction. The transition moments of the absorbing group then lie in a cone with semiangle ψ whose axis (the polymer chain) itself lies in a cone

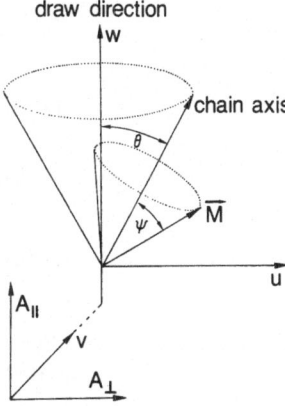

Fig. 6. Distribution of transition moments of a particular vibration in a uniaxially oriented polymer with respect to the draw direction (see text)

with semiangle θ and the draw direction as axis (Fig. 6). The expression for R then becomes

$$R = \frac{2\cot^2 \psi \cos^2 0 + \sin^2 \theta}{\cot^2 \psi \sin^2 \theta + (1 + \cos^2 \theta)/2} \tag{12}$$

Hence, if the direction of the transition moment with respect to the chain axis is known, the average orientation of the chain segments can be determined from the measured dichroic ratio.

From another point of view the orientation of a single polymer unit (e.g. crystallite or local segment in the amorphous phase) can be characterized by the three Eulerian angles which define the three rotations required to bring into coincidence a set of Cartesian axes in the structural unit with a reference set of Cartesian axes in the oriented polymer [55, 56]. The specification is considerably simplified for uniaxially oriented systems in which case the orientation distribution of structural units is random with respect to two of the Eulerian angles and is given by $F(\theta)$ only. Here, θ represents the angle between the chain axis within one unit and the draw direction. Most experimental techniques, however, do not yield $F(\theta)$ directly, but provide values for certain angular functions, for example $\langle \cos^n \theta \rangle$, averaged over the orientation of all structural units:

$$\langle \cos^n \theta \rangle = \frac{\int_{\theta=0}^{\pi} F(\theta) \cos^n \theta \sin \theta \, d\theta}{\int_{\theta=0}^{\pi} F(\theta) \sin \theta \, d\theta} \tag{13}$$

The orientation distribution function $F(\theta)$ can then be approximately reconstructed by a series:

$$F(\theta) = \sum_{n}^{\infty} \frac{2n + 1}{2} \langle P_n(\cos \theta) \rangle \, P_n(\cos \theta) \tag{14}$$

where the $P_n(\cos \theta)$ are *Legendre* polynomials of order n and the first three polynomials of even n read:

$$P_0(\cos \theta) = 1 \tag{15a}$$

$$P_2(\cos \theta) = \frac{1}{2}(3 \cos^2 \theta - 1) \tag{15b}$$

$$P_4(\cos \theta) = \frac{1}{8}(35 \cos^4 \theta - 30 \cos^2 \theta + 3) \tag{15c}$$

The averaged *Legendre* polynomials are usually designated as orientation functions and as far as infrared polarization measurements are concerned $\langle P_2(\cos \theta) \rangle$ is related to the dichroic ratio by the expression:

$$\langle P_2(\cos \theta) \rangle = \frac{(3\langle \cos^2 \theta \rangle - 1)}{2} = \frac{(R - 1)(R_0 + 2)}{(R + 2)(R_0 - 1)} = f \tag{16}$$

where R_0 has been defined in Eq. (9) and f is equal to *Fraser's* oriented fraction in Eq. (10) and also corresponds to *Hermans'* orientation function [64] defined for X-ray diffraction measurements. For parallel chain alignment f becomes unity, for perpendicular alignment $-1/2$, and for random orientation f becomes 0.

The primary advantage of this deduction is that the orientation functions determined by independent methods (e.g. X-ray diffraction, birefringence, sonic modulus) may be coupled with the results derived from IR dichroic measurements of absorption bands which are representative of the appropriate phase [58]. For example, selected infrared absorption bands from the crystalline region of an uniaxially oriented polymer can be measured and the dichroic ratios determined. The orientation function for the crystalline region of the polymer film, f_c, can be determined quantitatively from azimuthal wide angle X-ray diffraction measurements of the appropriate reflecting planes of the crystal [65]. Substitution of the measured f_c and R values into Eq. (11) yields directly the value of R_0 and hence the transition moment angle ψ for that absorption frequency. A higher degree of reliability in the experimentally determined transition moment angle can be achieved if samples with different degrees of molecular orientation are examined by both the infrared dichroic and X-ray diffraction methods. Equation (11) predicts that a plot of f_c, the crystal orientation function determined from X-ray diffraction measurements, versus $(R - 1)/(R + 2)$, will be linear with zero intercept. A least square evaluation of the data from this line should yield a value of the slope, $(R_0 + 2)/(R_0 - 1)$, and a quantitative value of the transition moment angle can then be calculated from the resulting R_0 value using Eq. (9) [66]. Provided the polymer under investigation can be treated as a two-phase system in a further step, the orientation functions of the amorphous (f_{am}) and crystalline (f_c) regions can be expressed in terms of an average orientation function f_{av} weighted by the amount of each phase present [67]:

$$f_{av} = \eta f_c + (1 - \eta) f_{am} \tag{17}$$

where η and $1 - \eta$ are the volume fractions of the crystalline and amorphous material, respectively. However, in view of the rather complex nature of orientation phenomena in polymeric systems, the analysis of anisotropy in terms of this relationship can only be treated as an approximation of the real situation.

5 Deuteration

Among isotope exchange techniques direct deuteration with gaseous or liquid D_2O in combination with IR spectroscopic investigations has been widely applied in studies of polymeric structure [6, 53, 68, 69].

Owing to the mass dependence of the vibrational frequency isotopic substitution reactions will result in frequency shifts of particular absorption bands of the polymer. The technique, however, is limited to polymers containing loosely bonded protons in NH or OH functional groups (e.g. cellulose, polyvinyl alcohol, polyamides, polyurethanes) and its applicability to the characterization of molecular order in polymers is based on the fact that hydroxyl and amino groups residing in chain segments of different degree of order are differently exposed to isotopic substitution. Thus, the rate and extent of the exchange reaction will strongly depend on the mechanical and thermal history of the polymer under investigation.

Given the applicability of Beer's law and assuming that the reason for a proton not to exchange is due to it being inaccessible to D_2O the percentage Z of accessible regions at any stage of deuteration can be determined from the relationship:

$$Z = \left[1 - \frac{A_{(X-H)}}{A_{(X-H)i}} \right] 100 \, (\%) \tag{18}$$

where $A_{(X-H)}$ is the absorbance of an absorption band belonging to a primarily uncoupled X—H vibration and the subscript i refers to the undeuterated sample.

The experimental procedure and the detailed construction of the deuteration cell used for such investigations have been reported previously [69]. Basically, the sample is mounted in a sealed cell and deuterated by bubbling a stream of dry nitrogen through liquid D_2O prior to flushing the cell which is kept at a constant temperature. The progress of isotope substitution in the film sample under examination can then be quantitatively monitored by the integrated absorbance of a particular absorption band which is characteristic of the extent of exchanged protons [e.g. $\nu(NH)$, $\nu(OH)$].

As an example the IR spectra of an undeuterated and a partially (40 %) deuterated polyester urethane film are shown in Fig. 7. The most obvious spectral changes upon deuteration directly reflect those absorption bands which belong to vibrations involving the NH group. Thus, the H—D isotope exchange results in partial replacement of the $\nu(NH)$ absorption band of the hydrogenbonded NH groups at 3331 cm^{-1} and absorption bands at 3180 cm^{-1} and 3115 cm^{-1} [70] by a band complex at about 2480 cm^{-1} [71]. While the absorption bands at 1703 cm^{-1} and 1733 cm^{-1}, essentially attributed to the $\nu(C=O)$ absorption of the hydrogenbonded urethane and non-bonded ester carbonyl groups, are almost unaffected by deuteration. The bands at 1590 cm^{-1} and 1531 cm^{-1} which have been interpreted as $\delta(NH)$ and $\delta(NH) + \nu(CN)$ modes [70-72] decrease in intensity upon substitution of H by D.

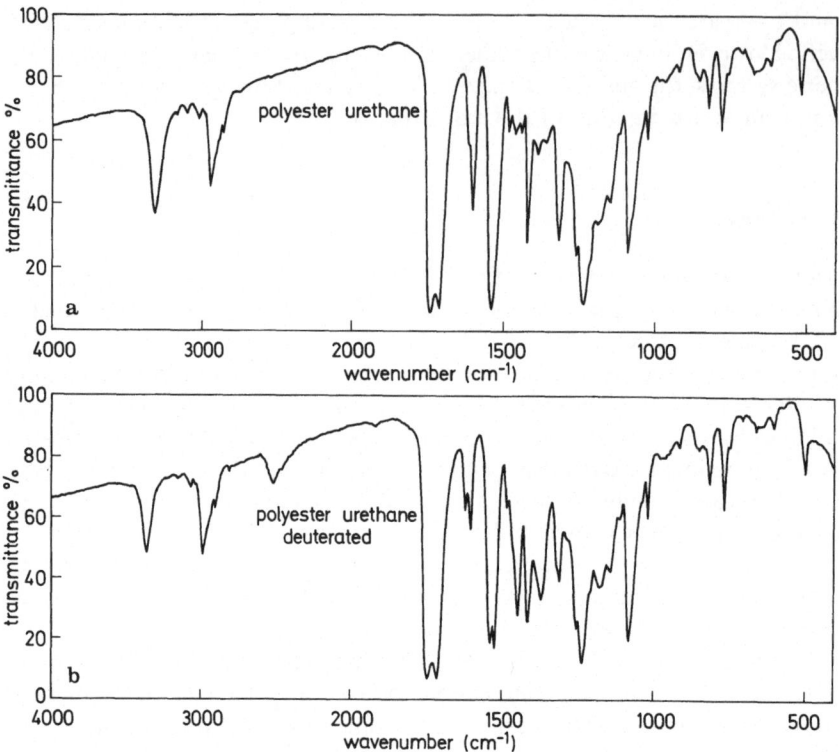

Fig. 7a and b. IR spectra of an undeuterated **a** and a partially ($\sim 40\%$) NH-deuterated **b** polyester urethane film

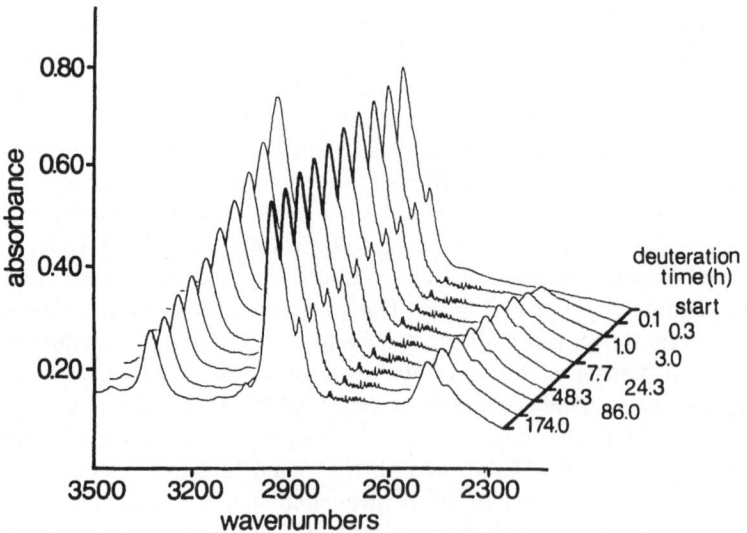

Fig. 8. FTIR spectra of a polyester urethane film in the 3500–2300 cm^{-1} wavenumber region taken in different time intervals during deuteration

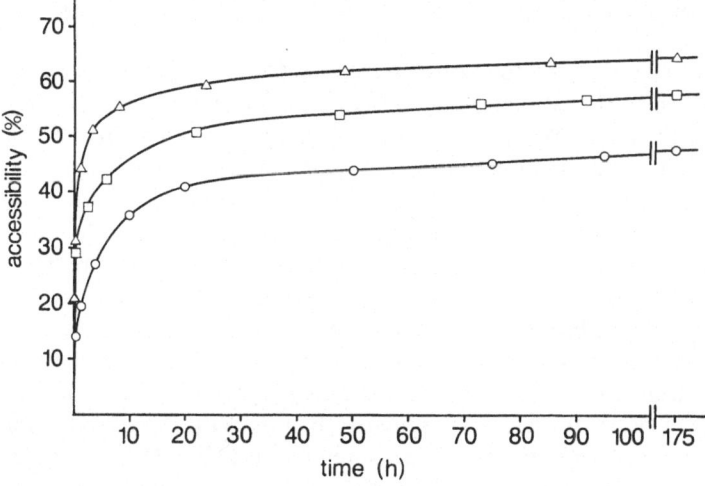

Fig. 9. Accessibility of three polyester urethane films (see Fig. 45) with different hard: soft segment composition as a function of deuteration time (—△— (a), —□— (b), —○— (c))

The time dependence of the isotope exchange reaction can be monitored quantitatively by taking IR spectra of the polyester urethane film in regular time intervals during the deuteration progress (Fig. 8) and evaluating the integrated absorbance of the v(NH) band with the aid of Eq. (18). The accessibility curves derived in this manner for three different polyester urethanes with increasing hard:soft segment ratio (see Chapter 7.2.2) are shown in Fig. 9. Principally, the curves can be separated into two portions. Thus, in the region of rapid exchange progress approximately 41, 50, and 59%, respectively, of the available NH protons have been substituted by deuterium within 20 hours. Previous investigations of aliphatic and aromatic polyamides of different degree of crystallinity [6, 73)] have established a direct correlation between the extent of readily deuterated protons and the percentage of amorphous regions determined by independent methods. In polyurethanes however, the accessible protons rather reflect the percentage of urethane groups of small hard segments dispersed in the soft-segment matrix while the region of reduced substitution rate corresponds to the slow penetration of the deuteration agent into larger, phase-separated hard segments [74)] (see also Fig. 44). In the samples under examination the different accessibility levels (Fig. 9) clearly demonstrate the larger proportion of inaccessible phase-separated hard segments with increasing hard-segment content. This hypothesis is further supported by about 5–10% higher accessibility values obtained in deuteration experiments on the same samples in the 200% drawn state. This result can readily be interpreted in terms of the disruption of larger hard-segment units upon drawing (see also Chapter 7.2.2).

In a separate chapter it will be shown that for polyurethanes a combination of the deuteration technique and rheo-optical FTIR investigations provide a more detailed understanding of phase-separation and the orientation behaviour of the individual domains during uniaxial elongation and recovery.

6 Variable Temperature Measurements and Hydrogen Bonding

Spectroscopic studies at variable temperatures have become an important tool for the characterization of the physical structure of polymers. Especially in combination with thermoanalytical DTA or DSC measurements short-time spectroscopic FTIR investigations in controlled heating or cooling experiments provide a detailed picture of the structural changes as a function of temperature. Any variations of spectroscopic parameters such as intensity, wavenumber position and band shape directly reflect the temperature dependence of the vibrational behaviour of the investigated polymer as a consequence of changes in the inter- and intramolecular interactions and the state of order [6, 75-80]. The vibrational spectra of polymers recorded in selected temperature intervals are of special value in studies of melting and recrystallization processes, thermal degradation, hydrogen bonding and polymorphism and greatly facilitate the assignment of conformational regularity bands. With the introduction of a variable-temperature cell for rheo-optical measurements (see Fig. 1) the potential of this technique has been further extended to study the temperature dependence of the mechanical properties of polymers.

As an example the FTIR spectra of semicrystalline poly(tetramethylene terephthalate) (PTMT) taken during heating (8.5 K/min) up to the melting point in 4.25 K (0.5 min) intervals are shown in Fig. 10. While the absorption bands associated with the vibrations of the aromatic ring and the carbonyl group [81, 82] show only relatively small changes of their integrated intensities during heating the absorption bands

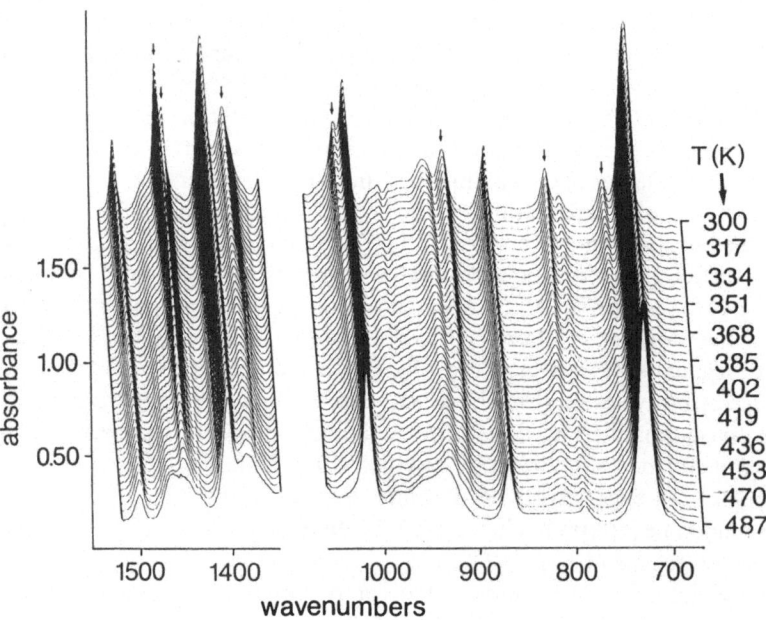

Fig. 10. FTIR spectra of semicrystalline PTMT taken during heating (8.5 K/min) in the 300–491 K temperature interval. The arrows mark the temperature-sensitive absorption bands

which can be assigned to CH_2-bending (1500–1400 cm^{-1}), CH_2-wagging (1400 to 1350 cm^{-1}) and coupled skeletal CH_2-rocking vibrations (1050–900 cm^{-1}, 850 to 750 cm^{-1}) of the g–t–$\bar{g}$ conformation of the aliphatic segments completely disappear at the melting point (491 K) of the polymer. As will be shown in chapter 7.1.3 these temperature-sensitive absorption bands are also most strongly affected by the stress-induced solid-state transformation observed in PTMT. This behaviour clearly reflects the important influence of the aliphatic segments on the thermal and mechanical properties of this class of polymers.

The role played by hydrogen bonds in the structure and properties of polymeric solids has been the subject of numerous investigations [6, 53, 83–87]. As far as the mechanical properties are concerned some investigators stress the importance of hydrogen bonding for these aspects [84, 85], while others conclude that too much emphasis has been placed on their role in determining mechanical properties [86, 87]. Although the energies of hydrogen bonds are weak (20–50 kJ/mol) in comparison to covalent bonds (of the order of 400 kJ/mol), this type of molecular interaction is large enough to produce appreciable frequency and intensity changes in the vibrational spectra of the examined compounds. In fact, the disturbances are so significant that IR and Raman spectroscopy provide the most informative source of criteria for the presence of hydrogen bonds [88].

Hydrogen bonding involves the interaction between a proton donating group $(R_1—X—H)$ and a proton acceptor $(Y—R_2)$ and may be described schematically by

$$R_1—X—H \cdots Y—R_2$$

The formation of a hydrogen bond is generally favoured by highly electronegative X and Y atoms with relatively small atomic radii (for example, O, N, F) and requires Y atoms with lone-pair electrons [88].

As a consequence of the hydrogen bonding forces the $v(XH)$ and $v(YR_2)$ stretching frequencies will be lowered, whereas the deformation frequencies associated with the motions of the H and Y atoms perpendicular to their X—H and Y—R_2 bonds, respectively, will be increased.

The energy of the hydrogen bond is directly reflected by the $v(XH \ldots Y)$-stretching and $\delta(XH \ldots Y)$-deformation vibrations. However, these vibrations are of extremely low frequency and relatively few reliable data are available [88, 89].

Most of the investigations so far reported deal with the observed frequency shift and intensity increase of the $v(XH)$-stretching vibration upon hydrogen bonding. Especially the frequency shift $\Delta v(XH)$ has been correlated with various chemical and physical properties of the hydrogen bond (e.g., enthalpy of formation, bond distance $R_{X \ldots Y}$) [88, 90]. With increasing hydrogen bond strength, for example, the $R_{X \ldots Y}$ distance decreases and this decrease is accompanied by increases in the difference between the associated $v(XH)$ and the nonassociated $v(XH)$ stretching frequency. In fact, for a number of different types of hydrogen bonds relationships between $\Delta v(XH)$ and R have been established [90]. Thus, for NH ... O bonds this relationship is expressed by the following equation:

$$\Delta \bar{v} = 0.548 \cdot 10^3 (3.21 - R) \tag{19}$$

where $\Delta \bar{v}$ is given in cm^{-1} and R in Å.

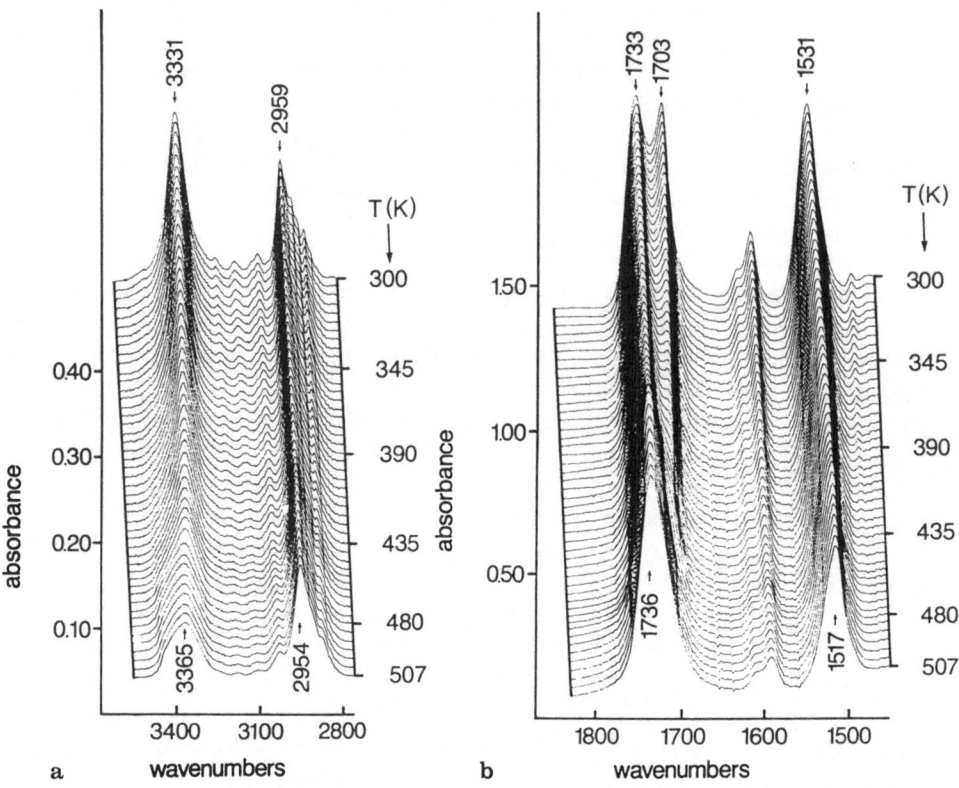

Fig. 11a and b. FTIR spectra of a polyester urethane film recorded in the 3500–2800 cm^{-1} and 1800–1400 cm^{-1} wavenumber region during heating at 9 K/min from 300 K to 507 K

The significant IR intensity enhancement which accompanies the formation of the hydrogen bond can be interpreted in terms of the augmentation of the X—H and electron donor dipole moment due to the charge redistribution produced by the new bond [91].

Dependent on the conformation of the polymer chains intra- or intermolecular hydrogen bonds may be encountered. Thus, the helix conformation of polypeptides and proteins is stabilized by intramolecular hydrogen bonds, while intermolecular hydrogen bonding occurs in the extended structure [92].

Hydrogen bonding is particularly important in polymers and copolymers containing the following functional groups: amide (polypeptides, proteins, polyamides), urethane (polyurethanes), hydroxyl (cellulose, polyvinyl alcohol), and carboxyl (polyacrylic acid).

IR spectroscopy has proved an excellent tool to study the hydrogen bonding behaviour in polyamides and polyurethanes. Thus, it has been shown that at room temperature over 99% of the NH protons in both even and odd members of the polyamide series [93] and about 80% of the NH protons in polyurethanes [70] are hydrogenbonded. Valuable information regarding the temperature dependence of hydrogen bonding may be derived from IR studies at elevated temperatures [6, 70, 74, 83, 94–97].

The 25-scan FTIR spectra of a polyester urethane film recorded in 30-second intervals during heating (9 K/min) from 300 K to 507 K (Fig. 11) exhibit characteristic temperature-dependent features for the $v(NH)$, $v(C=O)$, and $\delta(NH) + v(CN)$ absorption bands as a consequence of changes in the hydrogen bonding state. While the intense absorption band at 3331 cm^{-1} can be assigned to the $v(NH)$-stretching vibration of the NH groups associated through hydrogen bonds [$v_{ass}(NH)$], the small shoulder at about 3440 cm^{-1} is characteristic of the $v(NH)$ vibration of the non-hydrogenbonded NH groups [$v_{free}(NH)$]. Similarly, the $v(C=O)$ band complex can be separated in a $v_{ass}(C=O)$ band at 1703 cm^{-1} (primarily contributed from hydrogenbonded urethane carbonyl groups) and a $v_{free}(C=O)$ band at 1733 cm^{-1} which can be predominantly assigned to nonbonded ester carbonyl groups. With increasing temperature the following spectral changes are observed:

1. The intensity of the $v_{free}(NH)$ band increases at the cost of the $v_{ass}(NH)$ band.
2. The peak maximum of the $v_{ass}(NH)$ band is shifted toward larger wavenumbers.
3. The half width of the $v_{ass}(NH)$ band increases considerably with increasing temperature.
4. The $v_{ass}(C=O)$ band is shifted toward larger wavenumbers and eventually coalesces with the $v_{free}(C=O)$ band.
5. The $\delta(NH) + v(CN)$ band is shifted toward smaller wavenumbers.

The intensity decrease and increase of the $v_{ass}(NH)$ and $v_{free}(NH)$ absorptions, respectively, are indicative of the shift in equilibrium concentration of the hydrogen-bonded and non-hydrogenbonded NH groups. Unfortunately, these absorption bands overlap strongly and cannot be unambiguously resolved into their com-

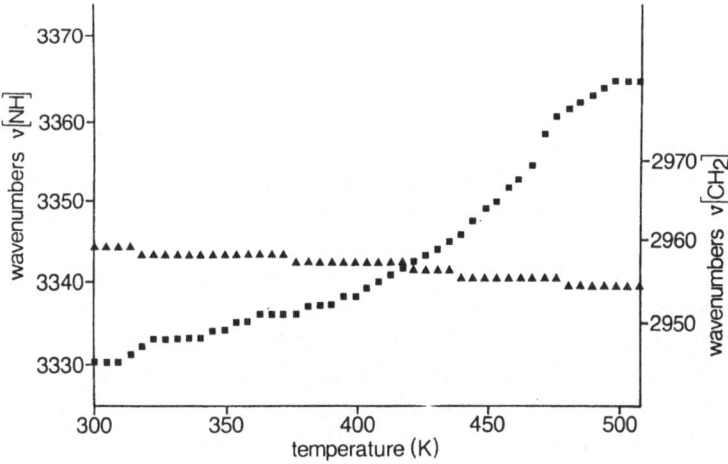

Fig. 12. Shift of the peak maximum wavenumber of the $v(NH)$ (■) and $v(CH_2)$ (▲) absorption bands of a polyester urethane as a function of temperature

ponents [96]. A total v(NH) absorbance area procedure for the quantitative assessment of hydrogen bonding in polyamides and polyurethanes has been proposed by Cooper et al. [70, 98]. The wavenumber shift and increase in bandwidth of the v_{ass}(NH) band at higher temperatures are the result of a general weakening of the hydrogen bonds and a concomitant broader distribution of their energies. Similar temperature-dependent spectral changes are also observed for the v_{ass}(C=O), absorption band. The frequency shift in the vibration of the bonded carbonyl groups, however, is less pronounced than that of NH groups because the acceptor atom is certainly less displaced than the hydrogen atom of the donor group [95]. The shift of the δ(NH) + + v(CN) band toward smaller wavenumbers reflects the predicted inverse effect of hydrogen bonding on stretching and deformation vibrations. In Fig. 12 the peak maximum wavenumbers of the v(NH) and v(CH$_2$) absorption bands have been plotted as a function of temperature with the aid of the automatic peak-search and plot routine. While the v(CH$_2$) peak maximum frequency shows only a slight linear decrease as a function of temperature drastic frequency shifts can be observed for the v(NH) absorption band in the 430 K–490 K temperature interval which corresponds to the melting range of this polymer (see Fig. 45c). With Eq. (19) a concomitant increase of $R_{N\ldots O}$ from 3.01 Å to 3.07 Å can be derived between 300 K and 500 K [99]. In view of the uncertainty of the exact wavenumber position of the v_{free}(NH) absorption band at about 3440 cm^{-1} the value of 3.01 Å for 300 K is in reasonable agreement with the value of 2.98 Å obtained by X-ray analysis on chemically equivalent model urethanes [100].

7 Experimental Results

7.1 Thermoplastics

7.1.1 Polyethylene

The infrared spectrum of polyethylene has been studied by numerous authors and a detailed band assignment is available for this polymer [6, 53, 101–103]. On the basis of these assignments several authors have attempted to monitor both statically and dynamically the orientation of the crystalline and amorphous regions as a function of strain [1, 4, 8, 104–109].

The band doublet observed in the spectrum of the semicrystalline polymer at 730/720 cm^{-1} can be assigned to in- and out-of-phase CH$_2$-rocking vibrations, respectively, of the crystal phase with the 730 cm^{-1} (B$_{3u}$) band polarized along the crystallographic a-axis and the 720 cm^{-1} (B$_{2u}$) band polarized along the b-axis. Under the assumptions that the contribution of the amorphous regions to the 720 cm^{-1} peak maximum absorbance is comparatively small with high density polyethylene [1] and that there is uniaxial symmetry within the sample we have attempted to monitor the orientation of the crystallographic a-, b-, and c-axes, respectively, relative to the stretching direction by their corresponding orientation functions f_a, f_b, and f_c on-line to the elongation procedure. From Eq. (11) it can be shown that f_a and f_b are related to the dichroic ratios of the 730 cm^{-1} and 720 cm^{-1} absorption bands by:

$$f_a = \frac{R_{730} - 1}{R_{730} + 2} \tag{20a}$$

and

$$f_b = \frac{R_{720} - 1}{R_{720} + 2} \tag{20b}$$

The orientation function for the crystál c-axis can then be evaluated from the relation [105]:

$$f_a + f_b + f_c = 0 \tag{20c}$$

The investigated polyethylene film (thickness 0.010 mm) was prepared by blown extrusion and had a density of 0.947 g cm^{-3}. In combination with DSC data a crystallinity of about 62% was determined.

The determination of crystal axes orientation in blown polyethylene films by wide-angle X-ray diffraction measurements has been extensively discussed by several authors [110-116]. Here, small- and wide-angle X-ray measurements [117] and IR polarization spectra of the untilted and tilted original specimens provided an approximate picture of the initial morphology and orientation of the crystallographic axes in the polyethylene film under examination. Thus, the b-axis is preferentially aligned parallel to the transverse direction of the film geometry while the a- and c-axes are distributed about the b-axis with the a-axis preferentially oriented in the machine direction [115, 117]. Only small differences in the orientation functions were derived from polarization spectra of the untilted specimens and specimens tilted about the machine direction for 45° in the undrawn and 400% drawn state. From the two-point small angle X-ray diagram ($\langle L \rangle = 22$ nm) of the original sample it can be concluded [117] that the superstructure consists of lamellae which extend predominantly in the transverse direction of the film specimen [113].

In different experiments film specimens were stretched in the machine direction at constant rate of elongation (85% strain per minute) at 300 K and 343 K and 10-scan spectra (resolution 4 cm^{-1}) were taken in about 11%-strain intervals with radiation polarized alternately parallel and perpendicular to the direction of stretch and unpolarized radiation. In Fig. 13 the FTIR spectra recorded during elongation at 300 K in the 690–750 cm^{-1} wavenumber region are shown separately for the parallel and perpendicular polarization directions alongside the corresponding stress-strain diagrams for 300 K and 343 K. The reduction of intermolecular forces in the polymer at elevated temperature is readily reflected by the considerably lower stress level of the 343 K experiment.

To correlate the macroscopic properties and the structural changes occurring during deformation the orientation functions evaluated from Eqs. (20a—c) have been plotted in dependence of strain in Fig. 14. Onogi and Asada [1] have discussed in detail the expected changes of infrared dichroism and orientation functions due to the different molecular processes of lamellar orientation and chain unfolding.

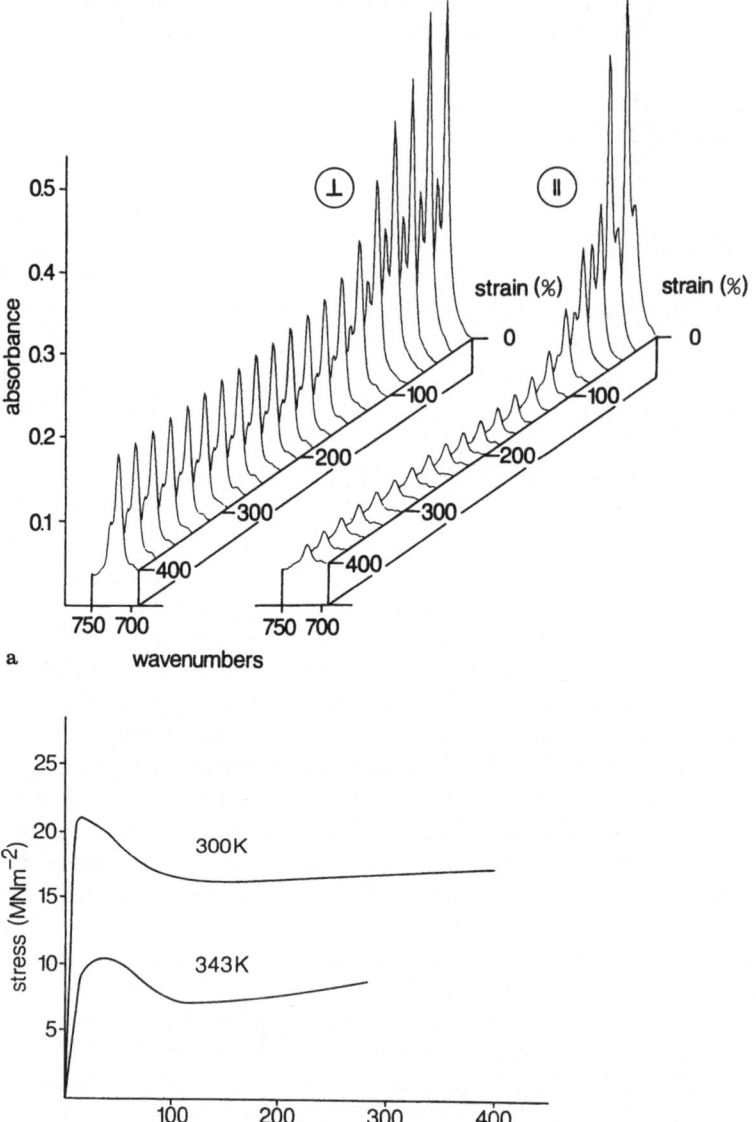

Fig. 13a. FTIR spectra taken at 300 K during elongation of a high-density polyethylene film with light polarized alternately parallel and perpendicular to the stretching direction; **b** Stress-strain diagrams of high-density polyethylene measured at 300 K and 343 K

Taking into account that unfolding of folded chains in polyethylene takes place when the lamellar structure is stretched in the direction parallel to the b-axis [118, 119] these authors have shown that for an original lamellar geometry with the b-axes oriented predominantly perpendicular and the a-axes oriented preferentially parallel to the subsequent direction of stretch (as it is the case in the film under investigation)

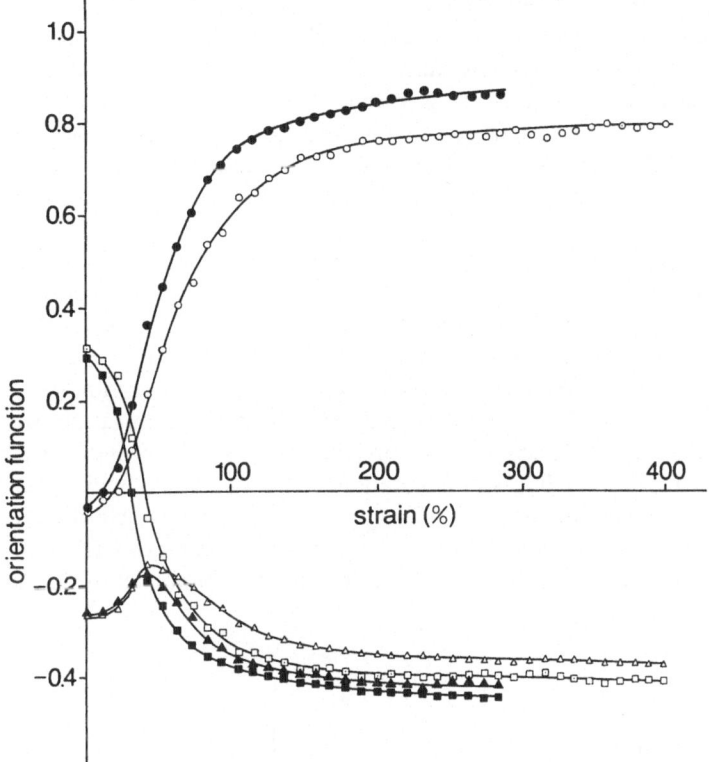

Fig. 14. The variations of the orientation functions f_a (□), f_b (△), and f_c (○) with strain for high-density polyethylene (open symbols: 300 K, closed symbols: 343 K)

orientational changes of the lamellar units as a whole induce an increase of f_b and a decrease of f_a whereas for the unfolding mechanism a decrease of both f_a and f_b can be predicted.

The orientation function-strain curve and the stress-strain diagram for uniaxial elongation at 300 K can be separated into different regions. The comparatively small orientation function changes in the elastic deformation region of the superstructural units below the yield point (0–15% strain) indicate that no appreciable change in lamellar orientation occurs in this region. From 15% to about 50% strain f_b and f_c exhibit different relative increases while f_a decreases drastically. This result corresponds to a rotational motion of the crystals with their b-axes and — more significantly — c-axes into the direction of stretch and their a-axes perpendicular to the direction of stretch. However, it should be kept in mind that it is difficult to differentiate between an orientational change caused by a rotation of a crystal and that caused by melting of a crystal and growth in a different state of orientation. The decreasing stress beyond the yield point is indicative of a partial disintegration of the superstructure into smaller blocks. The unfolding of molecular chains prevails at 50% strain where f_b passes through a maximum and subsequently both f_a and f_b show a decrease which levels off at about 150% strain where the applied stress reaches a minimum value after the yield point. In the region beyond 200% strain which is characterized by a weak increase of stress with strain only a further slight improvement of the orientation of recrystallizing unfolded chains takes place. Analogous deformation stages can be observed during elongation at 343 K with the only difference that a somewhat better orientation of the polymer chains in the direction of stretch is observed at higher strains. Thus, the orientation function value of 0.85 at 280% strain corresponds to an average inclination angle of the crystal c-axes with the draw direction of about 18°.

Apart from orientational changes the evaluation of the polarization spectra in terms of the structural absorbance [see Eq. (8)] of the 730 cm^{-1} absorption band provides a means to monitor the distortion and destruction of the orthorhombic crystalline modification during deformation [8, 53, 104, 120]. In the absence of another thickness reference band of appropriate intensity the 720 cm^{-1} absorption band was applied to eliminate the influence of decreasing thickness during elongation because an agreement within 5% between its peak maximum absorbance and the actual film-thickness reduction at 400% strain to about 40% of the original value was established. Therefore, the intensity ratio $A_{O_{730}}/A_{O_{720}}$ will directly reflect any consequences of the mechanical treatment on the perfection and proportion of the orthorhombic crystal structure. In Fig. 15 the structural absorbances of the 730 cm^{-1}

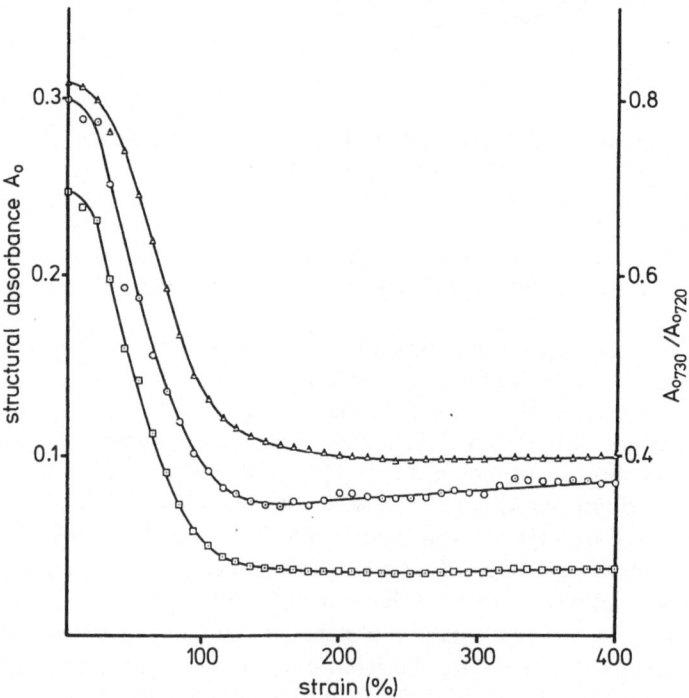

Fig. 15. Structural absorbances of the 730 cm^{-1} (—□—) and 720 cm^{-1} (—△—) band of high-density polyethylene and the corresponding ratio $A_{O_{730}}/A_{O_{720}}$ (—○—) evaluated for uniaxial elongation at 300 K

and 720 cm^{-1} absorption bands and the corresponding $A_{0_{730}}/A_{0_{720}}$ ratio are plotted as a function of strain. Beyond 20 % strain the absorbance ratio drastically falls off and reaches a minimum value of 0.34 at about 150 % strain. Between 150 % and 400 % strain a slight increase of this parameter indicates that obviously some recrystallization takes place in this strain interval. From the resultant smaller reduction of the $A_{0_{730}}/A_{0_{/20}}$ ratio observed during elongation at 343 K (a minimum value of 0.53 is attained at 150 % strain) it can be concluded that the mechanical treatment at elevated temperature has a less destructive influence on the orthorhombic modification of polyethylene. Although this approach does certainly not provide an exact quantitative picture of the "history" of the orthorhombic structure during the mechanical treatment some striking features can be detected by comparison with the corresponding stress-strain diagram. Thus, the onset of disruption of the orthorhombic crystal phase coincides with the yield point at 20 % strain and is subsequently accompanied by a substantial decrease of stress. The maximum extent of disruption according to Fig. 15 is reached at the lowest stress level beyond the yield point at about 150 % strain (Fig. 13 b). Further elongation finally leads to a reincrease of both stress and proportion of orthorhombic crystal phase.

The aforementioned structural changes occurring during uniaxial elongation at 300 K are additionally superimposed by a crystal phase transformation. Several authors have shown [6, 121, 122] that under conditions of stress the orthorhombic

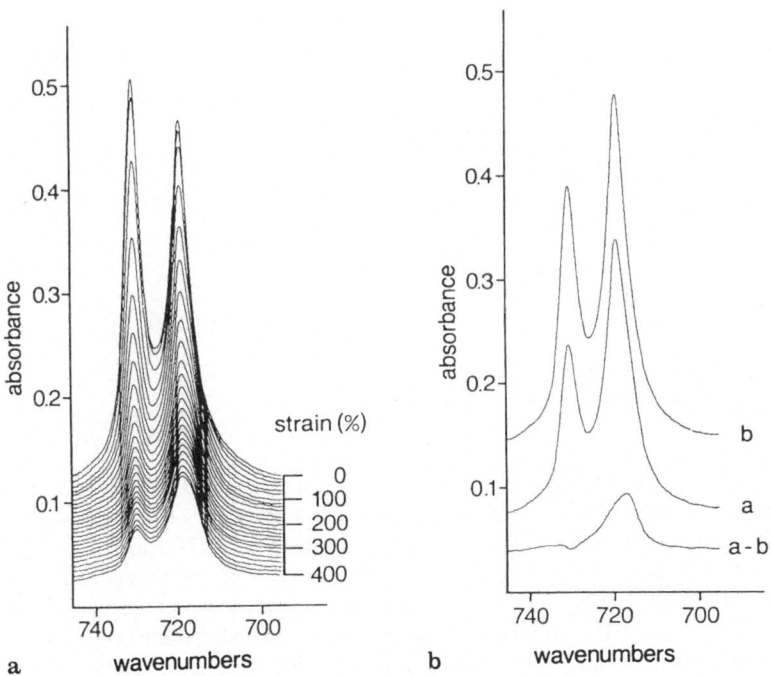

Fig. 16a and b. FTIR spectra recorded during uniaxial elongation of a high-density polyethylene film at 300 K with unpolarized radiation (**a**) and absorbance subtraction of successively recorded spectra (**b**) (a: 83 % strain, b: 50 % strain, a–b: difference spectrum)

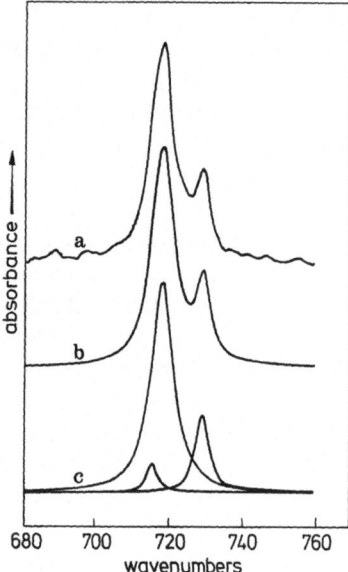

Fig. 17. Band separation of the $\gamma_r(CH_2)$ absorption bands: (a) experimental spectrum, (b) synthesized spectrum, (c) spectrum resolved in the individual absorption bands

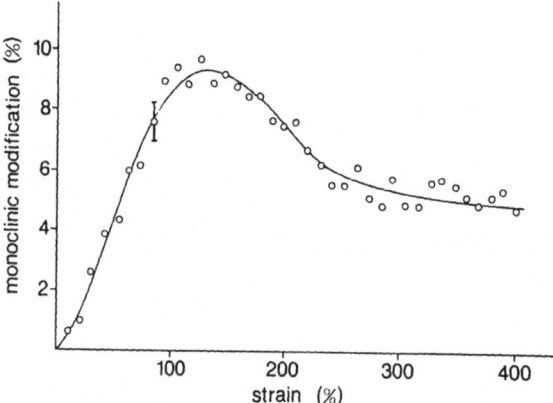

Fig. 18. Percentage of monoclinic modification of polyethylene derived from the $\gamma_r(CH_2)$ band complex of the spectra taken during uniaxial elongation to 400% strain at 300 K

modification of crystalline polyethylene is partially transformed to a monoclinic structure. The progress and the extent of this transformation in dependence of strain may also be studied with the aid of the CH_2-rocking mode. Unlike the 730/720 cm^{-1} band doublet of the orthorhombic modification the monoclinic structure is characterized by a single band near 717 cm^{-1} since the correlation splitting is expected to be small. Figure 16a shows the FTIR spectra of the polyethylene film under examination taken with unpolarized radiation and a resolution of 2 cm^{-1} during uniaxial elongation to 400% strain. The appearance of the 717 cm^{-1} shoulder can be readily accentuated by absorbance subtraction (Fig. 16b). For a more quantitative evaluation this absorption band was isolated by separation of the $\gamma_r(CH_2)$ band complex (Fig. 17) and the area under the isolated 717 cm^{-1} absorption band relative to the area of the total band complex was taken as representative of the proportion of transformed modification. The percentage of monoclinic modification derived in this manner for the individual spectra taken during elongation is plotted as a function of strain in Fig. 18. Interestingly, this curve runs through a maximum value of approximately 9% in the chain-unfolding region at about the same strain where the extent of disruption of orthorhombic modification reaches its maximum (Fig. 15). Finally, it should be mentioned that no significant transformation to the monoclinic modification could be observed for uniaxial elongation of the same sample at 343 K.

7.1.2 Isotactic Polypropylene

In one of the first attempts to apply rapid-scanning FTIR spectroscopy to the dynamic study of deformation phenomena in polymers isotactic polypropylene has been investigated by this technique [6, 7, 123].

Principally, in the IR spectrum of crystalline, isotactic polypropylene the absorption bands can be assigned to A- or E-mode vibrations with transition moment directions parallel and perpendicular to the 3_1-helix axis, respectively [101, 124–127].

The IR spectroscopic determination of the state of order and regularity in semicrystalline isotactic polypropylene has been discussed by numerous authors [58, 128–133]. For the quantitative characterization of chain alignment during uniaxial deformation the π-dichroic A-mode absorptions at 998 cm^{-1} and 973 cm^{-1} whose potential energy distributions are known in detail [127, 134] have been applied in the study under discussion. Thus, the 998 cm^{-1} band belongs to strongly coupled $\gamma_r(CH_3)$, $\nu(C—CH_3)$, $\delta(CH)$, and $\gamma_t(CH_2)$ vibrations while the 973 cm^{-1} absorption involves strongly coupled $\gamma_r(CH_3)$ and $\nu(C—C)$ backbone vibrations. In a FTIR absorbance subtraction study Painter et al. [135] have identified absorption bands which are characteristic of the regular 3_1-helical and irregular conformations in the ordered and amorphous phases, respectively. Hence, the 973 cm^{-1} absorption has been shown to contain contributions of both phases, while the 998 cm^{-1} band is predominantly representative of conformationally regular chains in the ordered domains.

The investigated isotactic polypropylene film samples were 20 mm in length, 5 mm in width and had a thickness of 0.030 mm and the density as measured in the gradient column was 0.905 g cm^{-3}. From this density a value of approximately 0.65 was derived for the volume fraction η of the crystalline regions. The molecular weight evaluated from intrinsic viscosity measurements on solutions in tetraline was about 90,000. Wide-angle X-ray studies revealed that the samples were in the smectic hexagonal modification. The polymer film was uniaxially drawn at 302 K with an elongation rate of 67% strain per minute and 8-scan spectra were recorded at 13.5%-strain intervals with a resolution of 4 cm^{-1} [123].

In Fig. 19a the FTIR polarization spectra taken during uniaxial elongation under the specified conditions are shown separately for the parallel and perpendicular polarization directions as a function of strain. With increasing elongation several absorption bands exhibit significant parallel and perpendicular dichroism. The

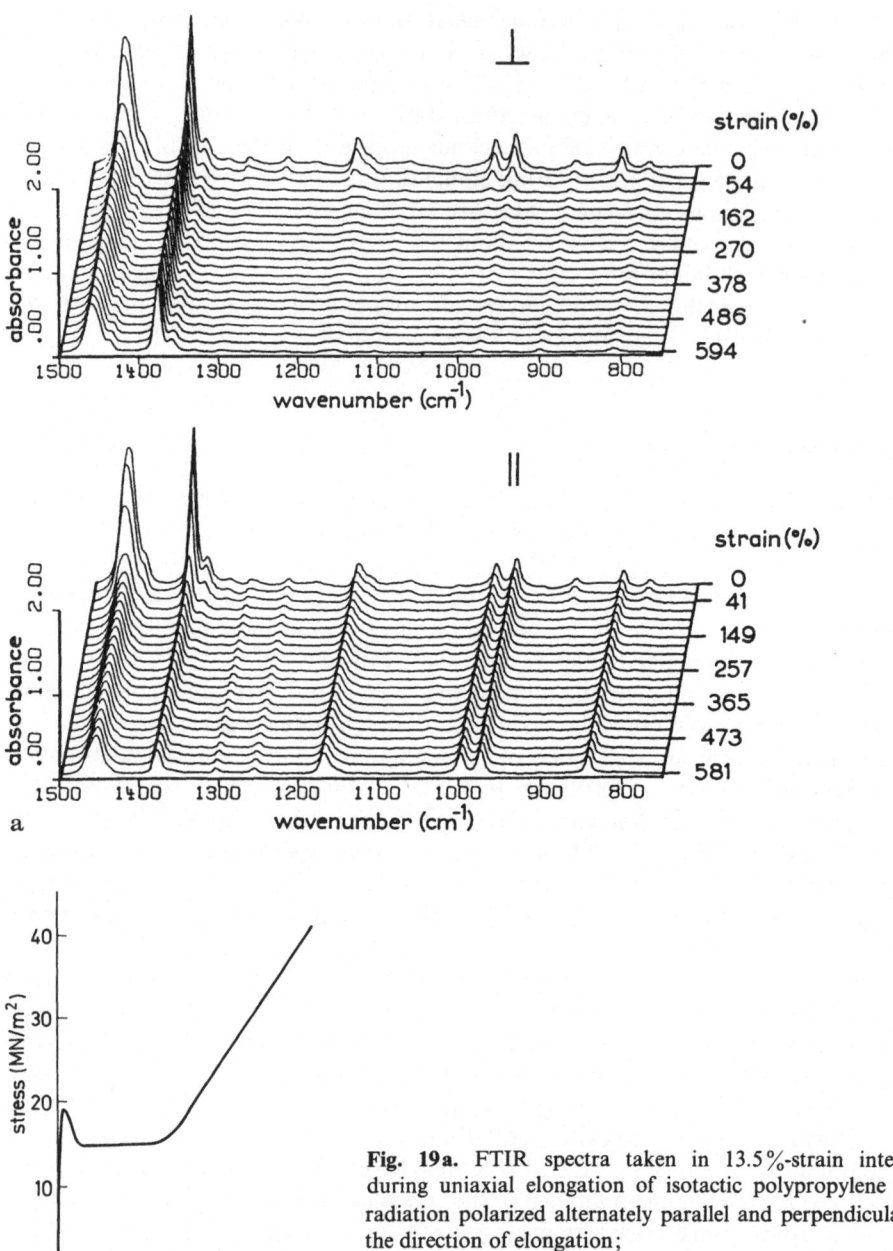

Fig. 19a. FTIR spectra taken in 13.5%-strain intervals during uniaxial elongation of isotactic polypropylene with radiation polarized alternately parallel and perpendicular to the direction of elongation;

b. Stress-strain diagram of isotactic polypropylene

stress-strain diagram corresponding to these polarization spectra is shown in Fig. 19b. The formation of a neck which propagates through the specimen is indicated by the yield point at about 20% strain and the subsequent plateau region up to 250% strain. Beyond 250% strain a linear increase of stress with strain is observed in the strain-hardening region. In Fig. 20 the dichroic ratios of the 998 cm^{-1} and 973 cm^{-1} absorption bands (determined from the integrated absorbances) have been plotted as functions of strain. In both profiles large π-dichroism effects are observed in the region from 50 % to about 150 % strain. These changes which are indicative of a preferential parallel alignment of the polymer helix axes both in the amorphous and crystalline regions with reference to the direction of stretch correspond to the propagation of the neck past the sampling area in the spectrometer. They may shift on the strain scale in different experiments dependent on the position of the initial formation of the neck relative to the sampling area. The dichroism measured in this strain range represents the average orientation of the inhomogeneous sample area exposed to the IR beam. Once the shoulder-neck region has completely moved past the sampling area, no significant dichroism changes are detected until the onset of the strain-hardening region. Beyond 250 % strain where the specimen has been reduced to uniform cross-section both the 973 cm^{-1} and 998 cm^{-1} bands reflect further π-dichroism increase which is more pronounced for the latter absorption. This may be interpreted by a preferential alignment of conformationally regular chains at this stage of elongation. In this context it seems interesting to mention that Takayanagi et al. [136] have reported in deformation studies of highly oriented polypropylene that above a draw ratio of $\lambda = 5$ the helix conformation of the taut tie molecules is partially deformed under the applied stress. In a very recent investigation Lenz [137] has discussed the coexistence of the regular and a distorted, irregular 3$_1$-helix conformation in drawn polypropylene in terms of the 998 cm^{-1} and 900 cm^{-1} absorption bands, respectively.

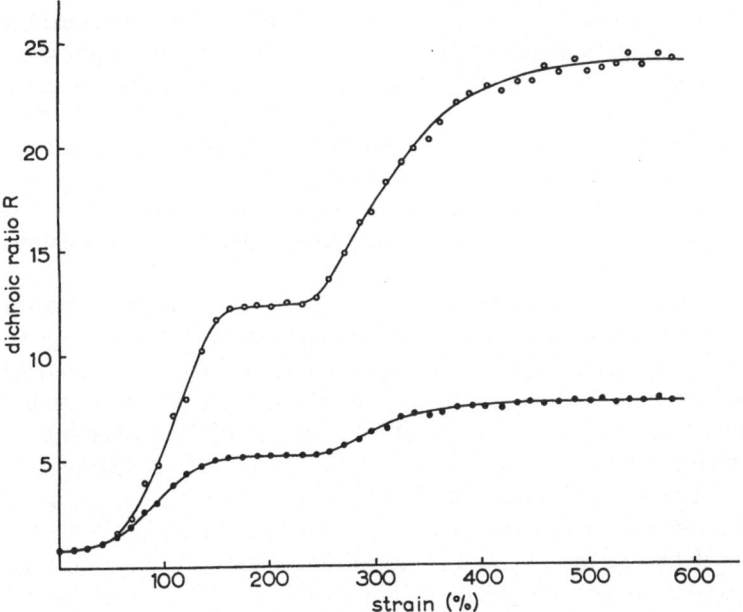

Fig. 20. Dichroic ratios R of the 973 cm^{-1} (●) and 998 cm^{-1} (○) absorption bands of isotactic polypropylene as a function of strain

Table 1. Orientation parameters of isotactic polypropylene derived from FTIR polarization spectra recorded during uniaxial deformation

Strain	Dichroic ratio		Orientation function		$\theta(°)$	
(%)	R_{998}	R_{973}	f_{998}	f_{973}	θ_{998}	θ_{973}
0	0.85	0.82	−0.05	−0.06	57	57
50	1.4	1.3	0.12	0.09	50	51
100	5.9	3.4	0.62	0.44	30	38
150	11.8	4.9	0.78	0.56	22	33
200	12.5	5.1	0.79	0.58	22	32
250	13.1	5.3	0.80	0.59	21	31
300	18.0	6.4	0.85	0.64	18	29
400	22.1	7.4	0.87	0.68	17	28
500	23.2	7.7	0.88	0.69	16	27
575	23.6	7.9	0.88	0.70	16	27

In Table 1 the dichroic ratios R, the orientation functions f, and the corresponding angles θ (see Eq. (12)) derived for the 998 cm^{-1} and 973 cm^{-1} absorption bands from the spectra of Fig. 19a are listed for selected elongation values. The initially negative orientation functions indicate that the original sample shows a slightly preferential orientation perpendicular to the direction of subsequent elongation. As pointed out earlier a larger decrease in angle θ during elongation has been observed for the 998 cm^{-1} band. Thus, a value of 16° was derived for, the 500% strained sample. From the azimuthal intensity distribution of the equatorial wide-angle X-ray reflection of a 500% drawn sample at relaxed stress level a value of about 8° was obtained for the average angle of disorientation of the polymer chains in the crystalline regions [138]. These experimental results strongly indicate that portions of conformationally regular sequences also occur in less oriented amorphous domains [139]. The polarization data of the 973 cm^{-1} absorption band which contains contributions both of the amorphous and crystalline phase yield significantly larger angles θ than the 998 cm^{-1} band and confirm that the polymer chains in the amorphous regions are on the average less regularly aligned in the direction of stretch than the polymer chains in the crystalline domains.

In Fig. 21 the orientation contributions of the amorphous and crystalline domains to the total average orientation have been separated and f_c (calculated from $R_{998\,cm^{-1}}$), f_{av} [calculated from $R_{973\,cm^{-1}}$) and f_{am} (derived from f_c and f_{av} with the aid of Eq. (17)] are plotted versus draw ratio for selected elongation values. The data are presented in logarithmic form so as to demonstrate the striking similarity with earlier studies of Samuels, in which samples from film and fiber high-temperature deformation processes were examined after fabrication by X-ray diffraction, birefringence, sonic modulus, and density measurements to obtain their orientation functions [66, 140, 141]. Thus, the linear correlation between f_{am} and the logarithm of the draw ratio has been interpreted in terms of the important role of the noncrystalline regions for the deformation above the glass-transition temperature [142, 143]. In analogy to the earlier results it was observed that a sharp change in slope occurred around a f_c of 0.8 to 0.9 at the point where a highly oriented microfibrillar structure appeared and

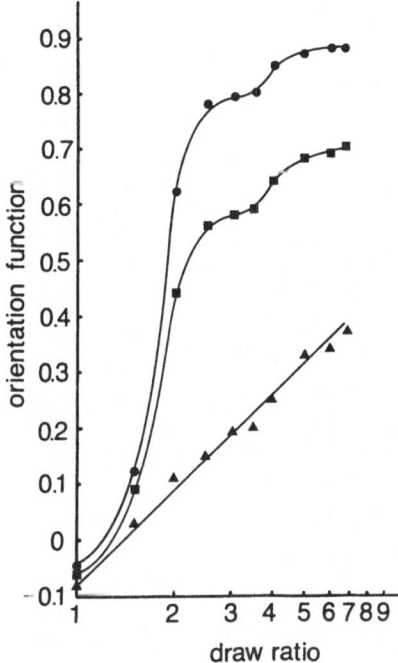

Fig. 21. Relation between the orientation functions f_{am} (▲), f_{av} (■), and f_c (●) and the draw ratio during uniaxial extension of an isotactic polypropylene film

further increases in orientation were eventually dominated by the noncrystalline regions. Due to a different assumption of the transition moment angle for the 973 cm^{-1} band the change in slope at the onset of the strain-hardening region at $\lambda = 4$ was found at a somewhat lower value of f_{av} (0.64) in our studies in comparison to Samuel's findings ($f_{av} = 0.75$). This agreement of structural behaviour in different deformation processes points toward general time-temperature superposition rules for the anisotropic response of semicrystalline polymers and demonstrates the availability of different process routes to equivalent states of order.

The results of recent rheo-optical FTIR investigations of the exceptional mechanical properties of hard-elastic polypropylene films are discussed separately in Chapter 7.2.3.

7.1.3 Poly(Tetramethylene Terephthalate)

In some polymers a reversible transformation from one conformation to another takes place under tension. Thus, upon application of stress the crystal structure of poly(tetramethylene terephthalate) (PTMT) changes from a crumpled near gauche-trans-gauche (g–t–ḡ) conformation of the aliphatic chain segments (α-form) to an extended form with an all-trans (t–t–t) sequence (β-form) (Fig. 22a) [144–156]. This solid state transformation between two triclinic crystal structures characteristically affects the mechanical properties of the polymer in the 5–15 % strain region (see Fig. 23). The concomitant increase of the fiber identity period from 1.16 nm to the almost fully extended repeat unit of 1.30 nm can be derived from the shift of the near-meridional ($\overline{1}04$) and ($\overline{1}06$) reflexes in the corresponding wide-angle X-ray diagrams (Fig. 22b).

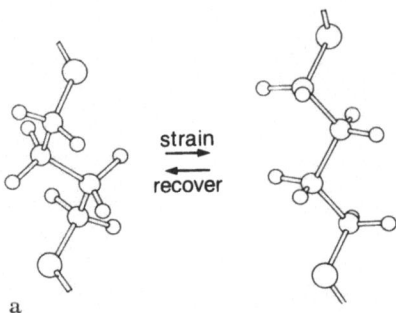

Fig. 22a and b. The stress-induced crystal phase transformation in PTMT: **a** conformation of the aliphatic chain segments in the relaxed and strained modification;

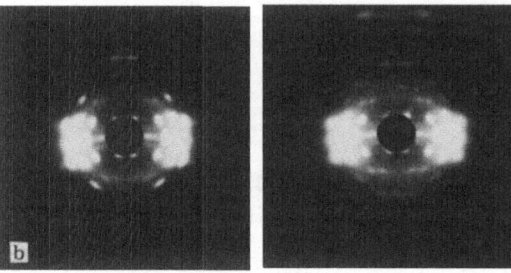

b wide angle X-ray diagrams of unstressed and stressed, slightly tilted PTMT fibers

The phenomenon of the PTMT crystal structure transformation can also be studied by infrared and Raman spectroscopy [6, 7, 81, 82, 149, 157–161]. Primarily the absorption bands associated with vibrations of the aliphatic segments in the relaxed and strained modification, respectively, may be utilized to quantitatively represent the reversible transformation as a function of the mechanical treatment. Hence, significant intensity enhancements with increasing strain are observed for the absorptions at 1485, 1471, 1393, 960 and 845 cm^{-1} while the bands at 1459, 1455, 1387, 916 and 810 cm^{-1} decrease with increasing strain and vice versa (see Fig. 24 and 25). The detailed assignment of these absorption bands to CH_2-bending, -wagging and -rocking vibrations coupled with skeletal modes has been discussed by several authors [81, 82, 161]. Siesler has first reported the application of rapid-scanning FTIR spectroscopy to monitor the spectral changes in small time intervals during elongation, recovery and stress-relaxation of PTMT films [7, 162–165].

The film samples utilized for the rheo-optical studies were prepared from PTMT chips by hot pressing at 523 K and quenching in ice water. Sections of this sample were drawn in hot water to a draw ratio of $\lambda = 4$ and finally annealed at constant length under vacuum at 483 K for 15 hours. The density of such samples was 1.328 g cm^{-3} and the film strips were cut 20 mm in length and 5 mm in width and had a thickness of approximately 0.025 mm.

A typical stress-strain diagram for a loading-unloading-loading cycle is shown in Fig. 23. The PTMT film under investigation has been extended at constant rate of 25 % per minute to 21 % strain (trace 1) returned to the original position (trace 2) and finally reextended until fracture occurred (trace 3). The stress-strain curve can be roughly separated into three portions. An initial steep rise of stress with strain

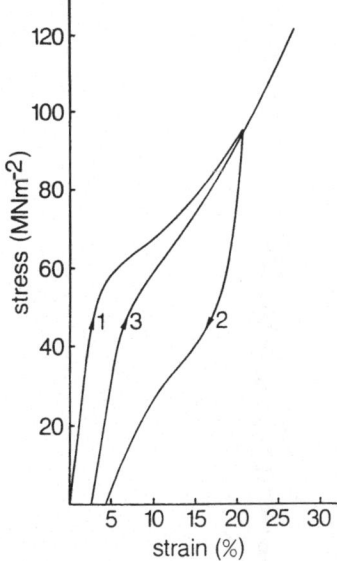

Fig. 23. Stress-strain diagram for a loading (1), unloading (2), and reloading (3) cycle of a PTMT film

is followed by a comparably flat portion between 5–15 % strain, and beyond 15 % strain the stress again rises steeply until fracture occurs [149, 150, 163]. Apart from a small permanent deformation (about 3 %) the induced strains are recoverable as shown by the unloading curve (trace 2) of Fig. 23.

The 4-scan FTIR spectra taken at 300 K with a resolution of 4 cm^{-1} in 1.24 %-strain intervals during the loading-unloading-loading procedure are shown in Fig. 24. They clearly reflect the significant intensity changes of the absorption bands associated with vibrations of the aliphatic segments [81, 82, 158, 161–163] as a function of strain. A more informative representation of the reversible transition is obtained by plotting the intensity of absorption bands which are characteristic of the stressed and relaxed segments, respectively, as a function of strain. For this purpose the absorbance variations of the 1459 cm^{-1} and 916 cm^{-1} (α-form) and 1485 cm^{-1} and 960 cm^{-1} (β-form) absorption bands were selected in Fig. 25. The absorbances have been corrected for changes in sample thickness with the aid of the aromatic $v(C{-}C)$ absorption at 1505 cm^{-1} as reference band.

Unlike detailed X-ray diffraction measurements [145, 148, 149, 153, 156] IR spectroscopic changes characteristic of the conformational transition can already be observed at about 3 % strain and extend beyond 20 % strain up to sample failure (see Fig. 25). A reasonable explanation for this discrepancy may be presented in terms of the microscopic behaviour of the material. While the occurrence of isolated stressed polymer units at low strains in the crystal lattice of the relaxed modification is directly reflected in the vibrational spectrum the appearance of the corresponding near-meridional wide-angle X-ray reflexes requires the spatial coherence of the scattering centers [152]. Owing to this requirement, the ($\bar{1}$04) and ($\bar{1}$06) reflexes of the stressed modification are first observed in the wide-angle X-ray diagram at distinctly higher strains than the IR spectroscopic changes. In analogy, unstressed segments are still detected in the IR spectrum beyond 20 % strain, although the ($\bar{1}$04) and ($\bar{1}$06) reflexes

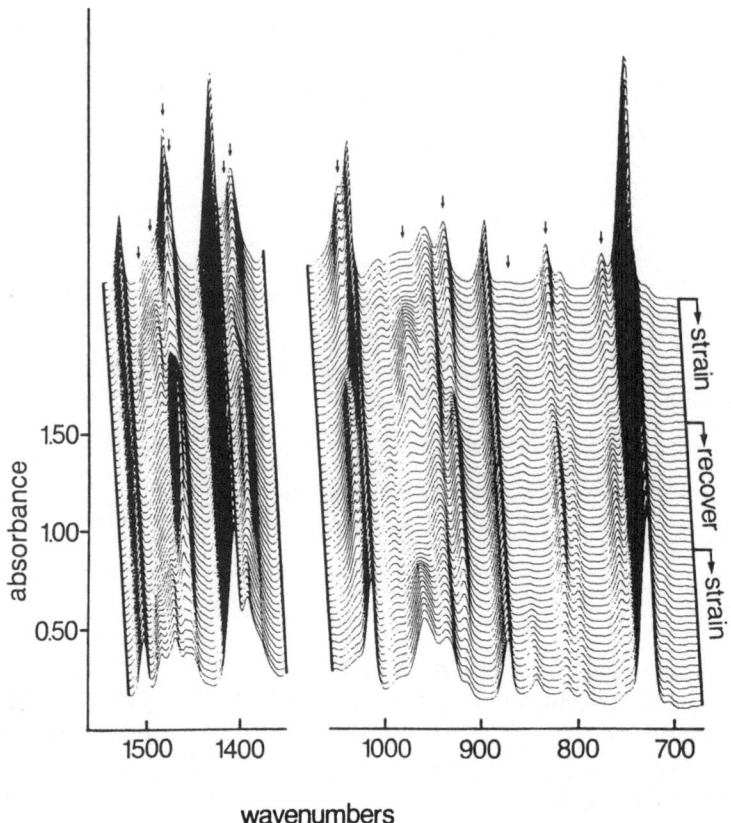

Fig. 24. FTIR spectra recorded during a loading-unloading-loading cycle of a PTMT film

of the relaxed crystal form have disappeared in the X-ray diagram [149, 163]. The percentage of residual unstressed modification in different samples as derived from the decrease of the 1459 cm^{-1} absorption band at strains of 20 % varied between 10 and 20 %. A reasonable explanation for this observation may be a nonuniform stress distribution in the samples as a consequence of nonuniformities in their thickness and in the perfection of their crystal structure.

The bend in the stress-strain diagram at about 3 % strain (Fig. 23, trace 1) coincides with the onset of the crystal phase transition as detected by FTIR spectroscopy and can be interpreted as a plasticization of the α-crystal lattice by the initial formation of elongated structural units. The equivalent, more diffuse bend in the reextension curve (Fig. 23, trace 3) is primarily a consequence of polymer segments which have been irreversibly elongated during the initial loading procedure. In the subsequent plateau region beyond 5 % strain both crystal forms are detected by X-ray diffraction and the most pronounced intensity changes occur in the FTIR spectra taken in this strain interval (Fig. 25). Here, the strain-optical coefficient $\partial A/\partial \varepsilon$ and the stress-optical coefficient $\partial A/\partial \sigma$ [167] (A = absorbance, ε = strain, σ = stress) of the conformation sensitive absorption bands of the α- and β-form reach their minimum

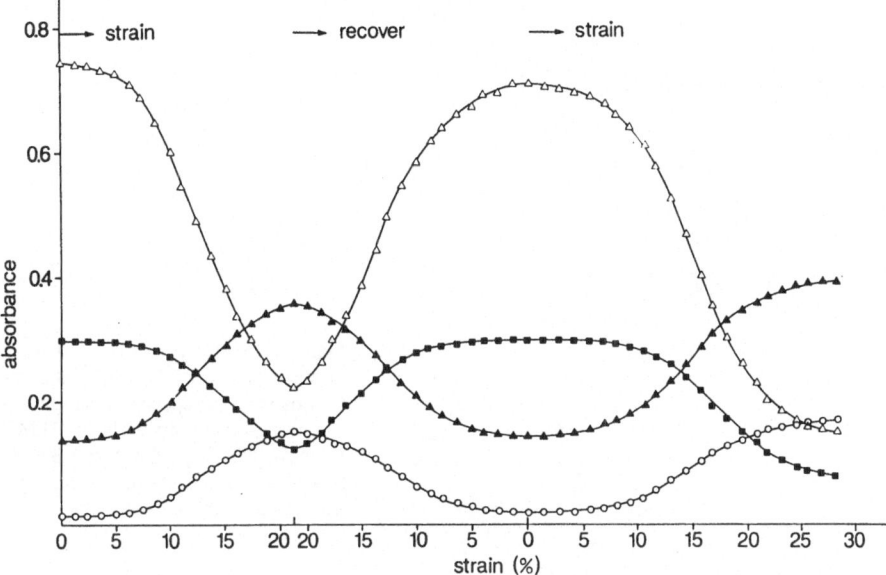

Fig. 25. Absorbances of the $\delta(CH_2)$ and $\gamma(CH_2)$ absorption bands as a function of strain: ($\triangle$) 1459 cm^{-1}, ($\blacksquare$) 916 cm^{-1} (α-form); ($\bigcirc$) 1485 cm^{-1}, ($\blacktriangle$) 960 cm^{-1} (β-form)

and maximum values, respectively, during elongation. Tadokoro et al. [158] correlated the critical stress σ^* for the transition from the crumpled α-form to the extended β-form with the stress at the maximum of the stress-optical coefficient of the β-form and confirmed by variable temperature measurements that σ^* corresponds well to the plateau region in the stress-strain diagram.

From the temperature dependence of the critical stress these authors derived the enthalpy and entropy differences between the α- and β-form:

$$\Delta H = H_\beta - H_\alpha = 5.1 \text{ kJ/mol of monomer unit} \tag{21a}$$

$$\Delta S = S_\beta - S_\alpha = 12.6 \text{ JK}^{-1}/\text{mol of monomer unit} \tag{21b}$$

The irreversible portion of the phase transition is not only observable in the stress-strain diagram but also becomes obvious in the spectroscopic measurements where the intensities of the absorption bands under examination do not completely revert to the initial values of the original sample upon unloading (Fig. 25).

A distinct temperature-dependent hysteresis effect is observable when the percentage of relaxed crystal α-form as determined from the absorbance of the $\delta(CH_2)$ absorption band at 1459 cm^{-1} is plotted for a loading-unloading cycle in dependence of the applied stress for 300 K and 353 K (Fig. 26). The characteristic values of the hysteresis loop widths of Fig. 26 [37 MNm^{-2} (300 K) and 25 MNm^{-2} (353 K)] are in good agreement with analogous, independent data obtained by Ward et al. [149] from wide-angle X-ray diffraction and Raman spectroscopic measurements. Furthermore, the critical stresses σ^* derived from Fig. 26 for elongation of PTMT films

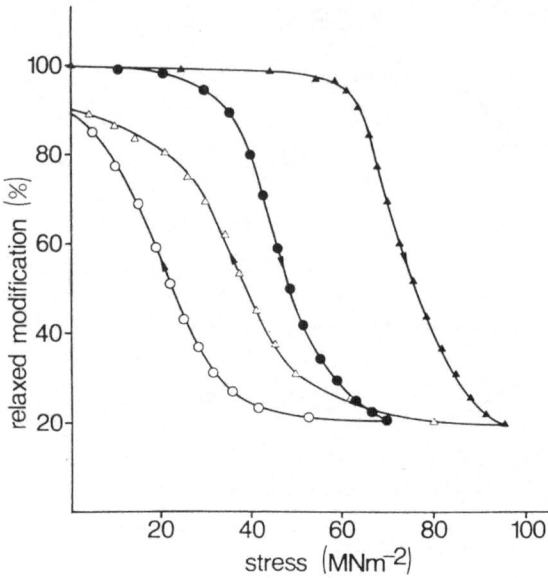

Fig. 26. Percentage of relaxed α-form derived from the 1459 cm^{-1} absorption band in the spectra taken during a loading-unloading cycle of PTMT films at 300 K (—△—) and 353 K (—○—): closed symbols (elongation), open symbols (recovery)

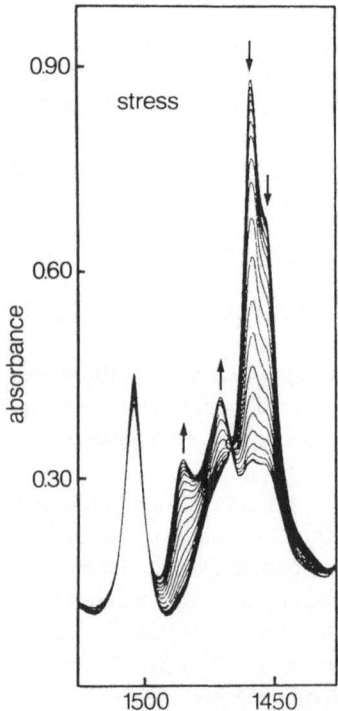

Fig. 27. Isosbestic point observed in the δ(CH$_2$) absorption region of FTIR spectra recorded during uniaxial extension of a PTMT film

at 300 K (72 MNm^{-2}) and 353 K (45 MNm^{-2}) are fully supported by the experimental results of Tadokoro et al. [158].

As an interesting detail of these investigations an isosbestic point characteristic of the equilibrium established between the relaxed and strained units at different stress levels could be detected in the region of the $\delta(CH_2)$ absorption bands (Fig. 27). This experimental result suggests that equivalent conversion of one modification into the other takes place with no possibility left for an intermediate variety forming [7, 149].

Contrary to annealed crystalline samples the stress-induced transformation of polymer chains in primarily amorphous PTMT to a conformation deviating slightly from the planar all-trans extended crystalline form has been found irreversible [165]. It has been shown that the spectra of strained crystalline and amorphous samples are very similar with the only difference that the higher frequency CH$_2$-bending mode of the elongated crystal form at 1485 cm^{-1} is shifted toward 1475 cm^{-1} in the strained amorphous specimen (Fig. 28). By inspection of the potential energy distribution [81, 166] in terms of force constants of the respective frequencies of the crumpled and elongated crystal modification one can deduce that the higher frequency band in both forms (1459 cm^{-1} and 1485 cm^{-1}, respectively) is mainly caused by interaction of the

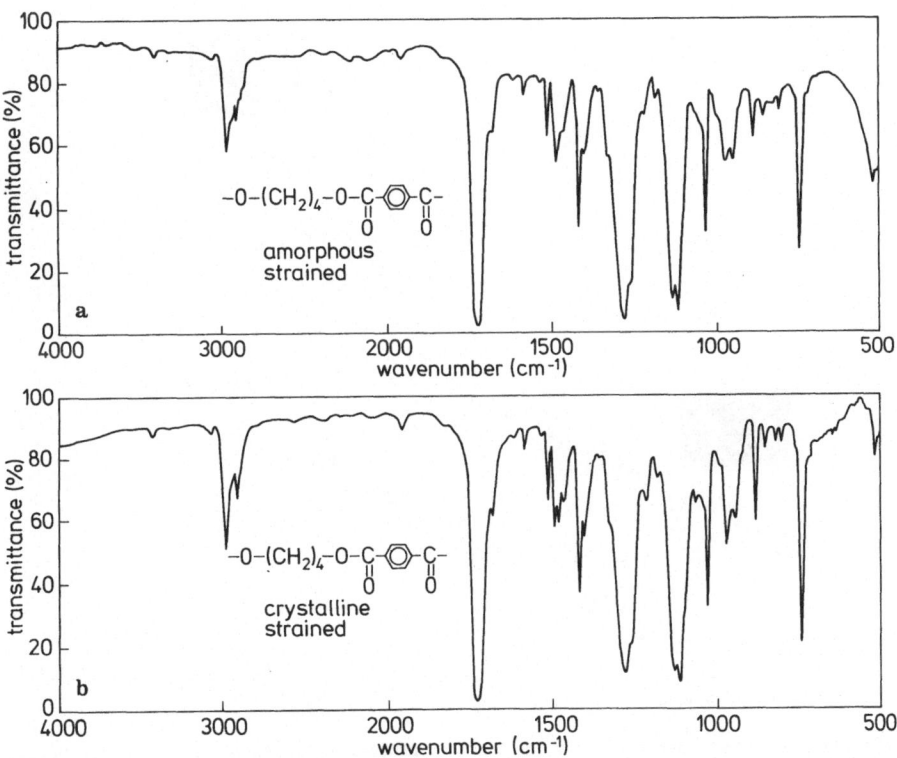

Fig. 28a and b. IR spectra of PTMT films: **a** oriented, strained, amorphous and **b** oriented, strained, crystalline sample

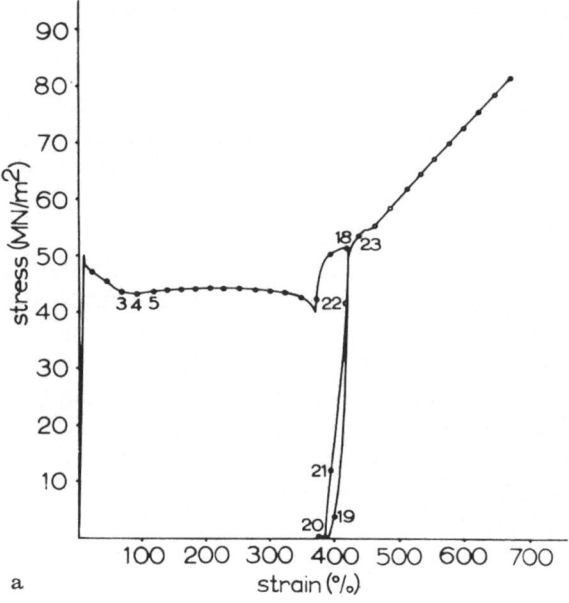

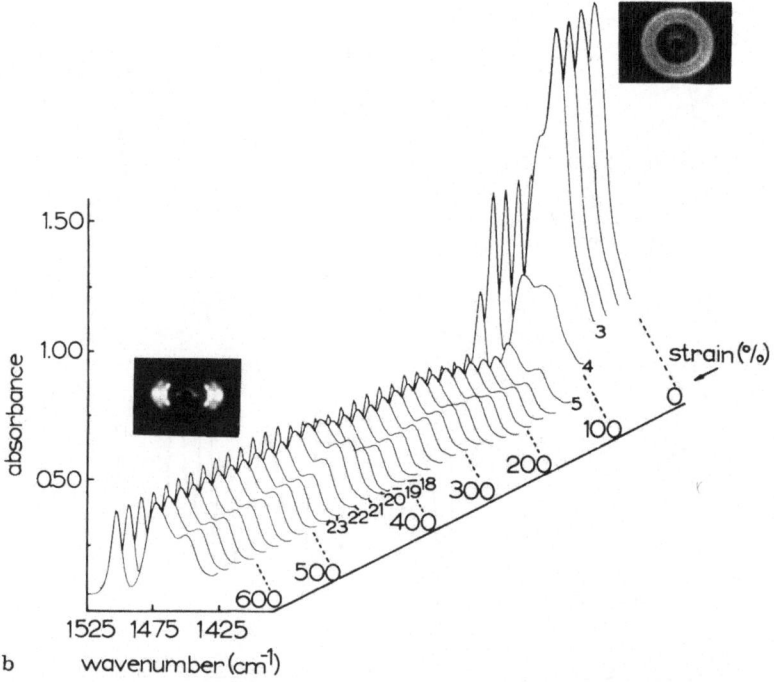

Fig. 29a and b. Simultaneous stress-strain and FTIR measurements during uniaxial deformation and recovery of a primarily amorphous PTMT film. **a** stress-strain curve of the mechanical treatment; **b** FTIR spectra in the 1500–1400 cm^{-1} $\delta(CH_2)$ region alongside the wide angle X-ray diagrams of the original and a 600% drawn sample

methylene groups adjacent to the oxygen atoms with their neighbourhood while the lower frequency mode (1455 cm^{-1} and 1470 cm^{-1}, respectively) can be predominantly assigned to vibrations of the central methylene groups in trans-conformation. The occurrence of the absorption band at 1485 cm^{-1} strongly depends on the planarity of the molecular skeleton [168]. Thus, it can be concluded that the shift of this band to 1475 cm^{-1} in the spectrum of the drawn amorphous specimen is caused by a conformation in which the CH_2-groups adjacent to the ester functionality deviate somewhat from the all-trans conformation of the almost planar, extended crystal form. Here too, a detailed picture of the response of amorphous PTMT film to the applied stress and any reversibility of the structural changes during uniaxial elongation and recovery in cyclic loading-unloading processes can be derived from simultaneous FTIR and mechanical measurements (Fig. 29). The film strips used for these investigations were cut 15 mm long and 10 mm wide from commercial PTMT film with a thickness of 0.040 mm and a density of 1.300 g cm^{-3}. The elongation rate was 100 % per minute and 12-scan spectra were taken at 300 K in 14-second intervals with a resolution of 4 cm^{-1}.

For the representation of the spectroscopic effects the most sensitive $1400-1500 \text{ cm}^{-1}$ $\delta(CH_2)$ region has been selected. As a further illustration of the orientation effects the wide-angle X-ray diagrams of the original sample and the 600 % elongated sample have been included in Fig. 29b. The stress-strain curve of the applied mechanical treatment is shown in Fig. 29a. The initial steep rise of stress with strain is followed by a plateau region between 50 % to about 350 % strain. In this region the neck whose formation starts at a threshold value of about 50 MNm^{-2} propagates through the entire film specimen with a concomitant reduction of sample thickness from 0.040 mm to about 0.015 mm. The spectroscopic changes corresponding to this necking region are illustrated in the spectra 3 to 5 (Fig. 29b) where the neck has been monitored with the laser reference system to move through the sampling area. Apart from a general intensity decrease as a consequence of the reduction in sample thickness the intensities of the $1459/1455 \text{ cm}^{-1}$ bands (not resolved here) of the crystalline α-modification drastically decrease while the intensity of the $1475/1470 \text{ cm}^{-1}$ band complex exhibits a relative increase. This result can be attributed to the predominant formation of a fibrillar structure with the abovementioned imperfect all-trans conformation of the aliphatic segments. Once the neck has completely moved across the specimen (at about 350 % strain) stress increases linearly with strain. To study the reversibility of the deformation the sample has been subjected to an unloading-loading cycle. The spectra taken during this cycle have been labeled 18–23 and the small intensity changes of the $1459/1455 \text{ cm}^{-1}$ bands relative to the $1475/1470 \text{ cm}^{-1}$ bands and the $v(C-C)$ aromatic ring reference band at 1505 cm^{-1} clearly reflect that there is only an extremely small portion of polymer chains reversibly recovering to the crumpled conformation upon unloading. This is also demonstrated in the wide-angle X-ray diagram of such an unloaded sample where the ($\overline{1}04$) reflex of the elongated crystal modification is still observable alongside the corresponding reflex of the relaxed α-form. These data lead to the conclusion that in predominantly amorphous specimens the reversibility of the conformational transitions occurring during elongation is strongly hampered. For the polymer chains in the amorphous domains this can be understood in terms of the lack of a driving force such as the improved molecular packing efficiency of the relaxed crystal form [152]. The loss of reversibility of the con-

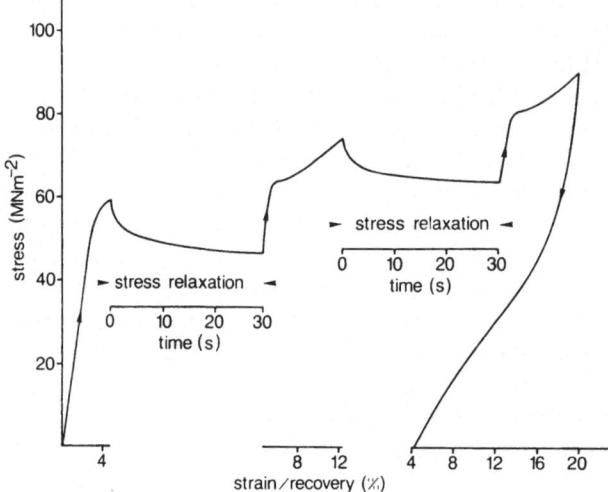

Fig. 30. Elongation and stress-relaxation of a semicrystalline PTMT film: stress-strain/time diagram

formational change for the small crystalline regions distributed in the amorphous domains may be readily explained by entanglement of the polymer chains in the amorphous matrix during the elongation process.

The elucidation of the molecular mechanisms responsible for the stress relaxation in semicrystalline polymers is a subject of considerable interest [169–173] and major advances in this field of research can certainly be achieved by the application of FTIR spectroscopy. Generally, the structural consequences of stress relaxation can be studied spectroscopically by focusing intensity changes of absorption bands which are sensitive to the state of order and orientation as a function of time after application of the stress.

Such measurements have been very recently applied with a new FTIR rapid-scan mode to characterize the time dependence of the conformational transition in PTMT during stress relaxation at different stress levels in the fast and slow decay region. Figure 30 shows the stress-strain/time diagram of a PTMT film sample (density 1.328 g cm^{-3}) which has been

a) initially extended to 5% strain

b) held at constant strain for 30 seconds

c) further extended to 12.5% strain

d) once again held at constant strain for 30 seconds

e) further extended up to 20% strain and

f) finally recovered to zero stress.

The elongation and recovery rate was 38% per minute and 1-scan spectra with a resolution of 4 cm^{-1} were taken in 0.35 second-intervals throughout the mechanical treatment. The intensity changes monitored in these extremely short time intervals during repeated elongation and stress-relaxation and final recovery at the peak maxima of the 1459 cm^{-1} and 960 cm^{-1} absorption bands which are characteristic of the crumpled and extended segments in the crystal phase, respectively, are shown in Fig. 31. Notwithstanding the short-time scans the FTIR spectra taken at specific

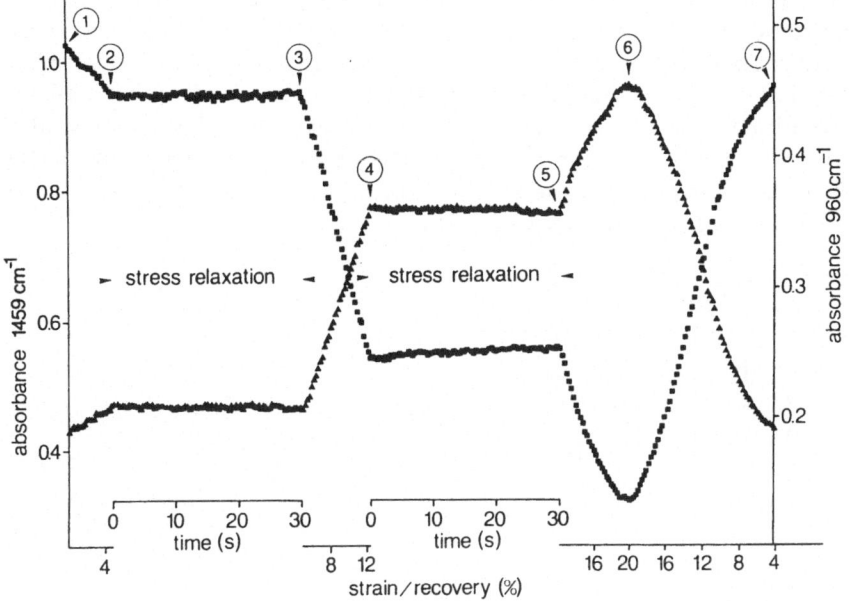

Fig. 31. Absorbance variations of the 1459 cm^{-1} (■) and 960 cm^{-1} (▲) absorption bands monitored during elongation, stress-relaxation and recovery of a PTMT-film (1–7 indicate those scans which are shown in detail in Fig. 32)

stress-levels during the mechanical treatment reflect an excellent signal-to-noise ratio (Fig. 32).

Despite a pronounced decrease of stress in the fast decay region no significant intensity changes of the conformation-sensitive absorption bands under examination can be detected during stress-relaxation at the lower stress level (60–50 MNm^{-2}). Contrary to previous investigations with a lower time resolution [164], however, a close inspection of Fig. 31 reveals that a slight transformation from the extended β-form to the relaxed α-form takes place during stress-relaxation at the stress level (75 to 65 MNm^{-2}) corresponding to the maximum of the stress-optical coefficient of the β-form (see above). The discrepancy can be readily explained in terms of the higher time resolution of the present spectroscopic measurements which allows a more accurate allocation of the absorption intensity corresponding to the onset of stress-relaxation. Nevertheless, the results indicate that only in the region of the critical stress σ* a small proportion of the extended crystal modification is reversed during stress-relaxation and an additional mechanism has to be taken into account. As the most probable mechanism which does not necessarily require a change in the proportion of relaxed and elongated structural units stress-relaxation can occur by slippage processes of intercrystalline segments. Depending on the shear-strength of the crystallites this mechanism may also involve the crystal phase [174]. According to the FTIR spectroscopic and mechanical results the process is characterized by an almost com-

plete conservation of the elastically stored deformation and a more uniform distribution of stress over the chain segments, respectively. The lastmentioned consequence
becomes evident in the stress-strain/time diagram upon reloading of the sample after
a certain period of stress-relaxation when the stress increases rapidly toward a level
which is significantly higher than at the onset of stress-relaxation. These findings also
support the view that chain rupture does not play an important role in the actual
stress-relaxation mechanism because fractured segments would certainly not contribute to an increase of the mechanical strength.

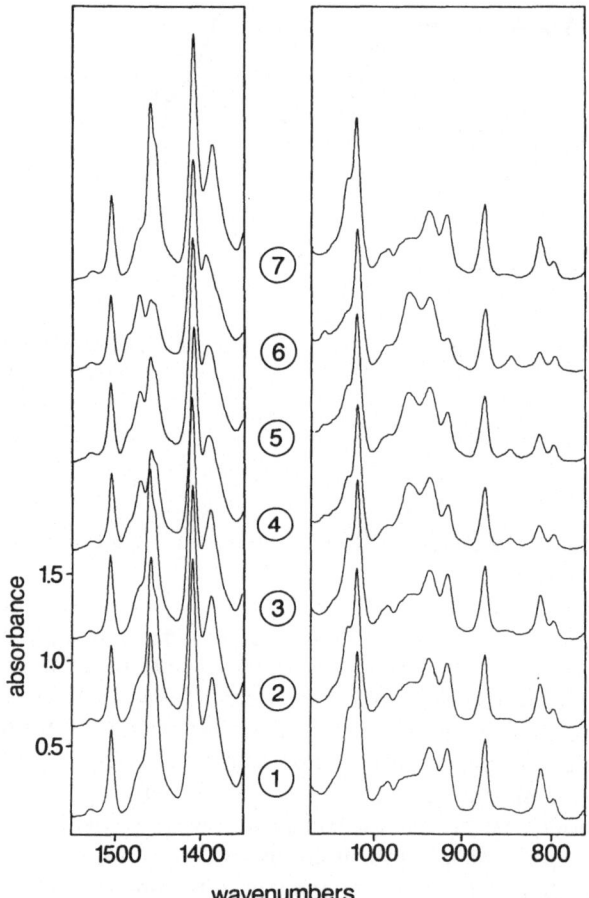

Fig. 32. FTIR spectra (0.35-second scans) monitored during the mechanical treatment shown in Fig. 30 (see also Fig. 31)

In conclusion, the experimental results favor a stress-relaxation mechanism in
semicrystalline PTMT which occurs primarily by slippage processes of chain segments
thereby homogenizing the transformation of forces between the crystallites by the
intercrystalline segments with only a small contribution from the reversal of extended
structural units to the crumpled α-form in the region of the critical stress.

7.2 Elastomers

7.2.1 Natural Rubber

Apart from the degree of crosslinking the mechanical properties of a polymer having network structure are strongly influenced by strain-induced crystallization [175–186]. This phenomenon is of great practical importance both during processing [181] and with regard to the technological properties of the product such as tear strength and maximum extensibility [182].

Elongation of a crosslinked elastomer decreases the entropy of the network chains and the additional decrease in entropy required for crystallization to occur is therefore relatively small. A schematic representation of strain-induced crystallization within a polymer network which has been elongated by a force in the specified direction is shown in Fig. 33. Crystallites thus formed act as crosslinks of high functionality

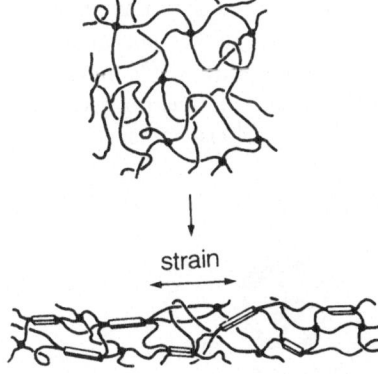

Fig. 33. Schematic drawing of strain-induced crystallization in a crosslinked elastomer

and since they are nondeformable at the stress levels involved, diminish the amount of elastomeric material able to respond to the imposed stress [175, 182]. Additionally, the crystallites act as filler particles which generally increase the modulus of a rubber-like material [182, 183]. In the stress-strain diagram strain-induced crystallization affects a drastic increase of elastic force with strain as a consequence of a significant self-reinforcement of the elastomer during elongation (see Fig. 34a) [175, 177, 179, 181–185]. The elastic force f is given by Eq. 22 in terms of the elongation ratio λ [175]:

$$f = \nu kTA(\lambda - \lambda^{-2}) \tag{22}$$

where ν and A are the density of network chains, i.e. their number per unit volume and the undeformed cross-sectional area, respectively. Contrary to numerous experimental results Eq. 22 implies that f should be proportional to λ at very large elongations. Such stress-strain data are customarily represented using the semi-empirical equation of Mooney and Rivlin [187–188]

$$[f^*] = 2C_1 + 2C_2\lambda^{-1} \quad . \tag{23}$$

in which C_1 and C_2 are constants independent of λ and the reduced stress or modulus [f*] is given by [189-191]

$$[f^*] = \frac{f}{A(\lambda - \lambda^{-2})} \tag{24}$$

A linear relationship between [f*] and λ^{-1}, however, holds only at small elongations and an upturn in the reduced stress occurs at small reciprocal elongations less than 0.4. The deviation from the linearity is controversely interpreted on the one hand by the limited chain extensibility [177, 192-194] and on the other hand by strain-induced crystallization [179, 180, 182].

In this context rheo-optical FTIR spectroscopy has proved a valuable technique to study the phenomenon of strain-induced crystallization on-line to the deformation process of the elastomer under investigation. Whith the aid of an appropriate absorption band which is characteristic of the threedimensional order in the crystalline phase the onset and progress of strain-induced crystallization during elongation and its disappearance upon recovery can be unambigously monitored simultaneously to the mechanical measurements. Representative for several rubber-like materials which have been investigated by this technique in our laboratory [195] the results obtained with sulfur-crosslinked (1.8 % S) natural rubber (100 % 1,4-cis-polyisoprene) and a radiation-crosslinked synthetic polyisoprene (93 % 1,4-cis-isomer) shall be discussed in some detail here.

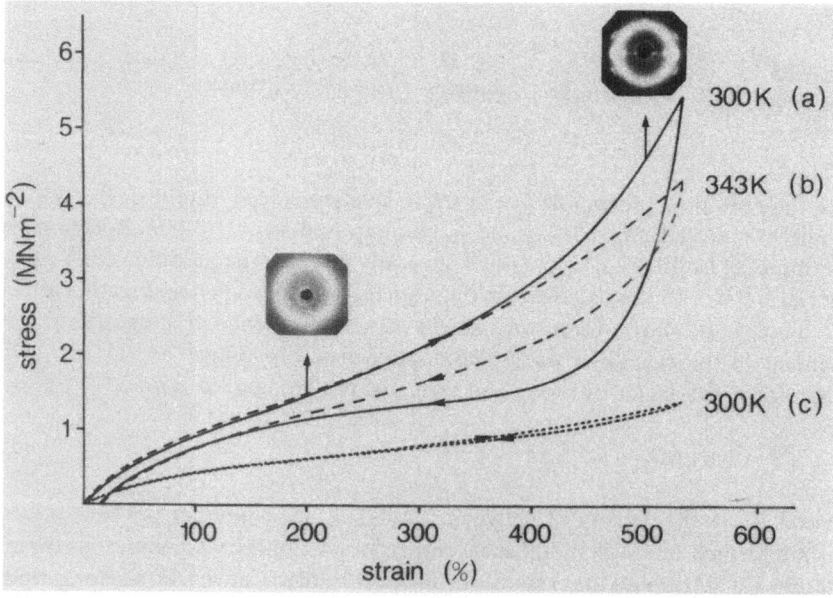

Fig. 34. Stress-strain diagrams of various crosslinked rubbers: (a) sulfur-crosslinked natural rubber (100 % 1,4-cis-polyisoprene) at 300 K, (b) sulfur-crosslinked natural rubber at 343 K, (c) radiation-crosslinked synthetic polyisoprene (93 % 1,4-cis-isomer) at 300 K (see text)

The samples were prepared by microtoming from sheets of the polymers film sections which were 4 mm in width, 15 mm in length and about 100 μm in thickness. In the mechanical treatment the films were subjected to a loading-unloading cycle up to 530% strain with an elongation and recovery rate of 85% strain per minute and 12- to 15-scan spectra were taken in 9- to 11-second intervals with a resolution of 4 cm^{-1}. In Fig. 34 the stress-strain diagrams of the sulfur-crosslinked natural rubber measured at 300 K (a) and 343 K (b) are shown alongside the corresponding

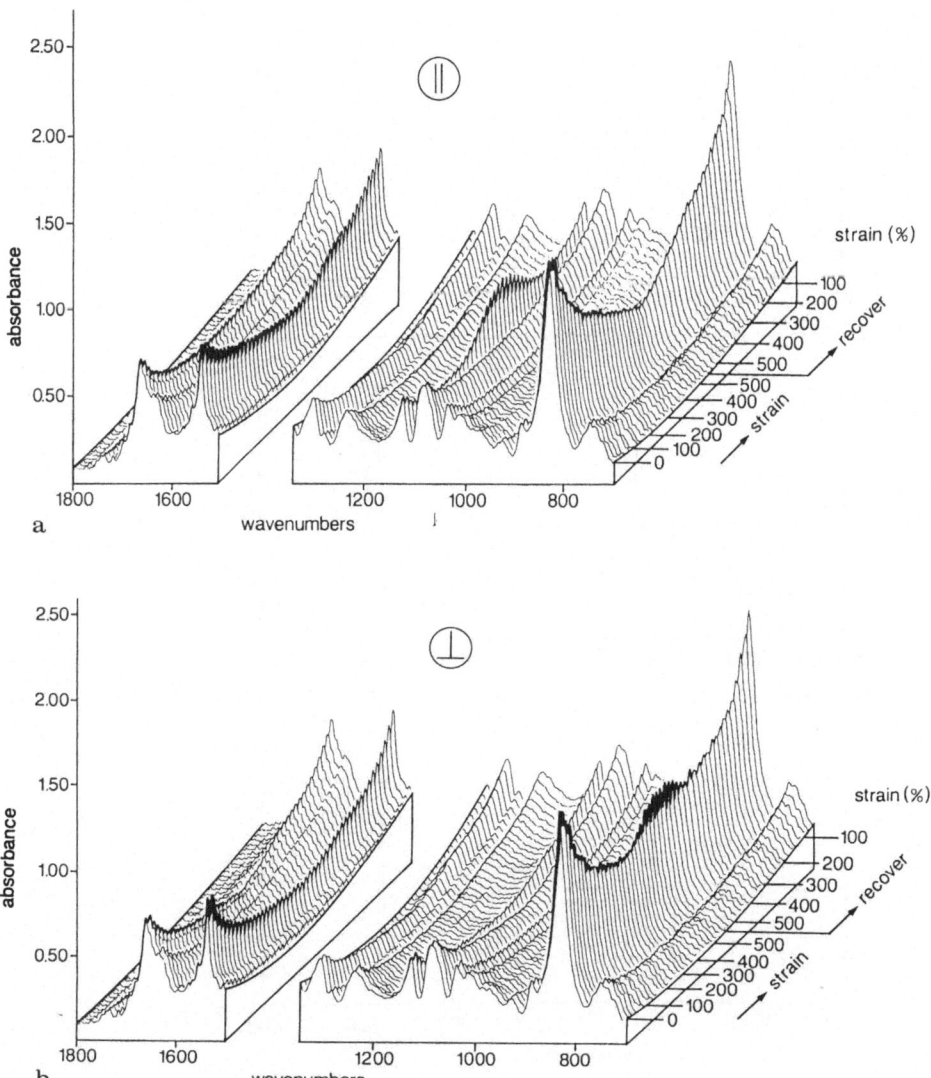

Fig. 35a and b. FTIR polarization spectra taken alternately with light polarized parallel **a** and perpendicular **b** to the stretching direction during an elongation-recovery cycle of a sulfur-crosslinked natural rubber at 300 K

data taken at 300 K with the radiation-crosslinked synthetic polyisoprene (c) which
did not crystallize during elongation primarily due to the configurational irregularity.
The increasing tendency for strain-induced crystallization from (c) to (a) is accompanied by a significant enhancement of stress hysteresis [7% (c), 21% (b), 36% (a)]
in the loading-unloading cycles of the different experiments. The hysteresis arises
from the tendency of the crystals formed on extension to persist as the tensile force is

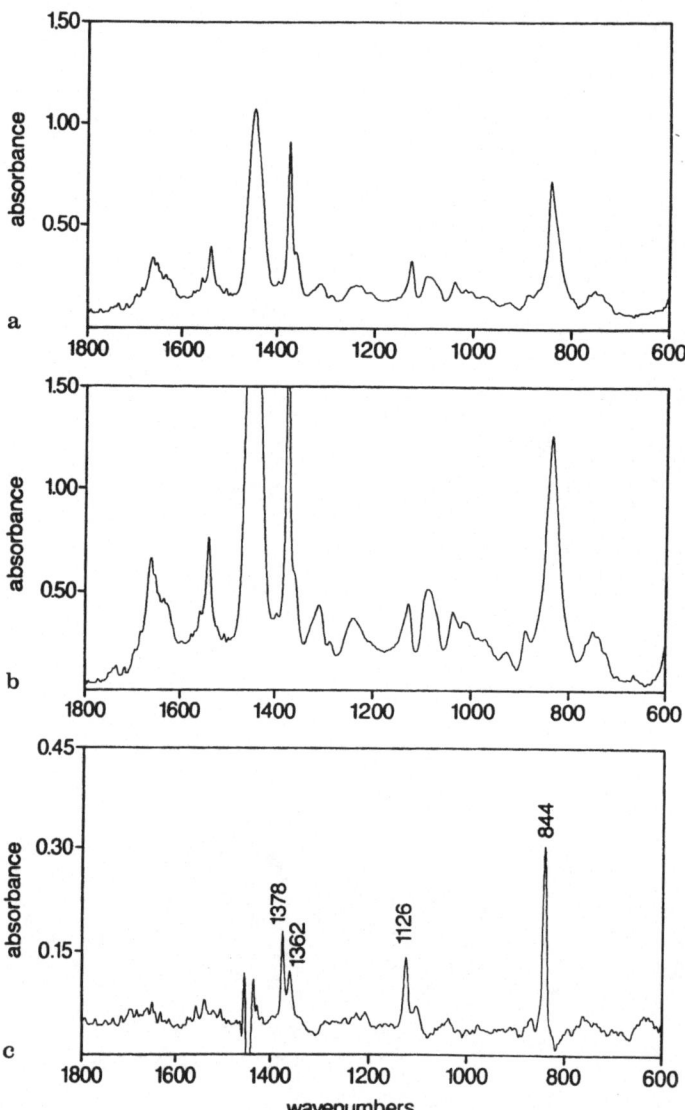

Fig. 36a–c. Accentuation of the strain-induced crystallinity bands in the spectrum of sulfur-crosslinked
natural rubber by absorbance subtraction: **a** structural absorbance spectrum of 500% elongated
sample; **b** structural absorbance spectrum of unstretched material; **c** difference spectrum **(a)–(b)**

reduced [192)] (see Fig. 37a). Ultimately, however, the amorphous state is recovered and the stress approaches the corresponding values in the loading half-cycle. In Fig. 34a the wide-angle X-ray diagrams taken at 200% and 500% strain, respectively, of the stress-relaxed samples have been included to demonstrate the change in the state of order during elongation.

The FTIR spectra monitored during such an elongation-recovery cycle of sulfur-crosslinked natural rubber at 300 K with light polarized alternately parallel and perpendicular to the direction of stretch are shown separately for the two polarization directions in Fig. 35. Although the actual absorption intensities are superimposed by dichroic effects in both spectra series absorption bands can be detected which increase in intensity despite reduction of sample thickness during elongation. For the accentuation of these crystallinity-sensitive bands the spectrum of the original sample has been subtracted from the spectrum of the 500% elongated sample with an appropriate scaling factor (Fig. 36). In order to eliminate the influence of orientation effects the absorbance subtraction was performed on the structural absorbance spectra synthesized from the individual polarization spectra according to Eq. (8).

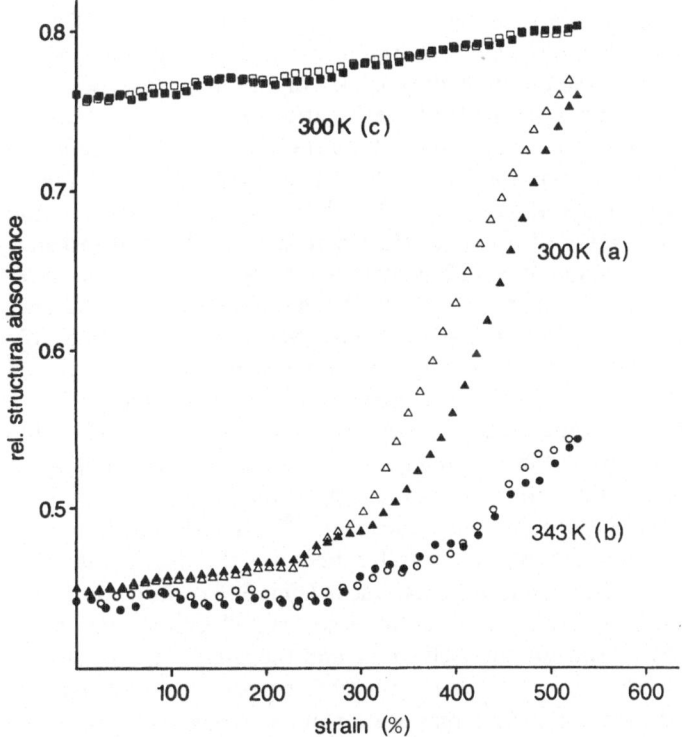

Fig. 37. Plot of the structural absorbance ratios $A_{0_{1126}}/A_{0_{1662}}$ versus strain: (a) sulfur-crosslinked natural rubber at 300 K, (b) sulfur-crosslinked natural rubber at 343 K, (c) radiation-crosslinked synthetic 1,4-cis-polyisoprene at 300 K (closed symbols: elongation, open symbols: recovery)

In the difference spectrum (Fig. 36c) at least four absorption bands at 1378, 1362, 1126 and 844 cm^{-1} have been isolated which can be associated with the crystalline phase formed during stretching. The discontinuity in the 1450 cm^{-1} wave-number region is caused by overabsorption of the $\delta(CH_2)$ and $\delta_{as}(CH_3)$ absorption bands in the spectrum of the unstrained sample.

The vibrational spectrum of polyisoprene has been discussed by several authors [196-204] but owing to the lack of data manipulation capabilities before the advent of FTIR spectroscopy discrepancies existed as far as the accentuation and the unambiguous assignment of the crystallinity bands was concerned.

To monitor the onset and extent of strain-induced crystallization during the loading procedure the 1126 cm^{-1} absorption band which has been assigned to a C—CH$_3$ in-plane deformation vibration [196, 199] and which shows the largest relative increase as a consequence of crystallization was utilized in the present study. An interesting behaviour is demonstrated by the C—H out-of-plane (835 cm^{-1}) absorption band [199, 203, 204] which shifts to 844 cm^{-1} (Fig. 36c) in the crystallizing specimen. In combination with polarization measurements it was therefore concluded [196] that the 835 cm^{-1} and 844 cm^{-1} absorption bands are representative of the amorphous and crystalline regions of 1,4-cis-polyisoprene, respectively. The $\nu(C=C)$ absorption band at 1662 cm^{-1} is almost perfectly compensated by the subtraction procedure (Fig. 36c) and has been applied as reference band to compensate the reduction of sample thickness during elongation. Thus, the ratio of the structural absorbances of the 1126 cm^{-1} and 1662 cm^{-1} bands was evaluated with the aforementioned software routine for the spectra taken during the loading-unloading cycles of the individual polymers (Fig. 34) and plotted as a function of strain for the elongation and recovery procedure in Fig. 37. The phenomenon of strain-induced crystallization is most clearly reflected by the data of curve (a) which correspond to the loading-unloading cycle of sulfur-crosslinked natural rubber at 300 K (Fig. 34a). Here, the onset of crystallization can be assigned to a strain value of about 230% where the structural absorbance ratio first deviates from the initial curve with a subsequent drastic increase at larger strains. Furthermore, a significant retention of the strain-induced crystallinity relative to the loading half-cycle is observable during recovery down to the threshold strain value of crystallization. Thus, basically, the process of crystallization and melting is reversible and can take place in the time scale of the experiment [205, 206]. In order to test the effect of temperature increase on strain-induced crystallization a sulfur-crosslinked natural rubber film specimen has been subjected to an analogous mechanical treatment at a temperature of 343 K, approximately 70 K above the melting point of the crystal phase in the unstressed state [186]. In agreement with the results of thermoelastic investigations of natural rubber [207] the applied stress is initially higher in the elevated-temperature experiment (Fig. 34b) and crosses over to lower values at elongations were the strain-induced crystallization commences at room-temperature. The reason for this behaviour and the smaller hysteresis can readily be derived from the corresponding structural absorbance/strain-plot in Fig. 37b. Thus, despite a comparatively larger scatter of the structural absorbance ratios due to a lower signal-to-noise ratio at elevated temperature the spectroscopic data clearly reflect the much lower degree of strain-induced crystallization at 343 K and the shift of its onset toward higher strains. In actual fact, strain-induced crystallization has been detected by rheo-optical FTIR measurements up to 373 K [208].

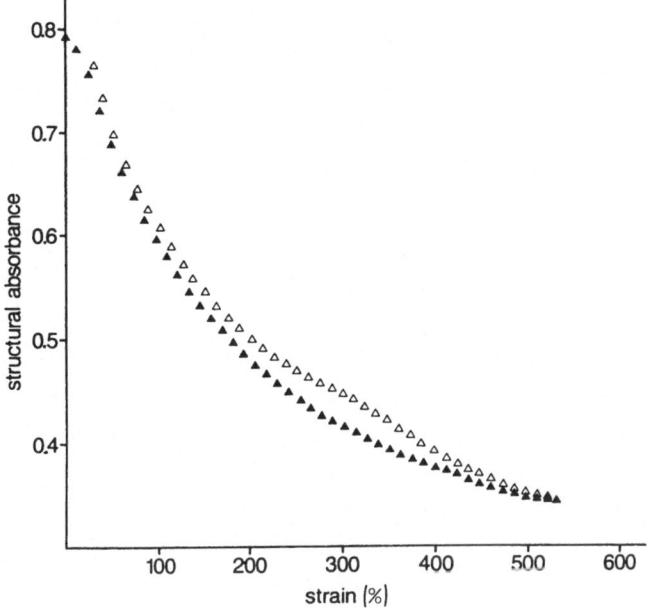

Fig. 38. Structural absorbance/strain-plot of the 1662 cm^{-1} v(C=C) reference band of sulfur-cross-linked natural rubber monitored at 300 K (closed symbols: elongation, open symbols: recovery)

When the chains in the amorphous network are stretched out because of the applied deformation the entropy of fusion is significantly diminished. Due to the inverse relationship to the entropy the melting temperature of the crystal phase is thereby increased and crystallization is induced in some of the network chains even at temperatures far above the melting point of the unstressed sample [209]. As expected from the stress-strain diagram no crystallization can be detected in the radiation-crosslinked synthetic 1,4-cis-polyisoprene (Fig. 37c). The relative shift of the structural absorbance ratios to higher values is merely due to a different baseline evaluation of the reference band for this polymer.

An interesting effect becomes evident when the structural absorbance of the v(C=C) thickness reference band is plotted as a function of strain (Fig. 38) for the loading-unloading cycle of the strongly crystallizing system (Fig. 34a). Thus, a distinct asymmetry in the recovery of the original sample geometry could be detected whenever the polymer under examination exhibited extensive strain-induced crystallization. The phenomenon may therefore be correlated with the composite nature of such a strain-crystallized system and must be interpreted in terms of a preferential recovery of the thickness relative to the transverse dimension in the stress-strain plateau region of the unloading half-cycle between about 400% and 200% strain (Fig. 34a).

The assumption that the 1126 cm^{-1} absorption band is characteristic of the three-dimensional order or conformational regularity of the crystal phase and not associated with perfectly elongated individual chains only, is supported by orientation measurements at elevated temperature [208]. Thus, despite a drastic reduction of strain-induced crystallization the degree of polymer chain orientation evaluated in terms of the

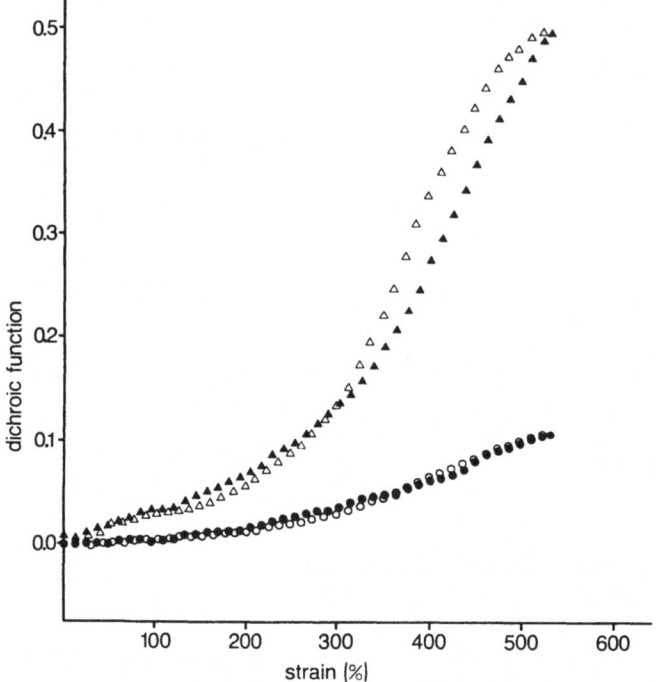

Fig. 39. Dichroic function/strain-plot of the 1662 cm^{-1} (◯) and 1126 cm^{-1} (△) absorption bands corresponding to the loading-unloading cycle of sulfur-crosslinked natural rubber at 300 K (closed symbols: elongation, open symbols: recovery)

dichroic ratio of the $v(C=C)$ absorption band at 1662 cm^{-1} did not significantly deviate from the measurements at ambient temperature.

To monitor the crystalline and amorphous orientation of a strongly crystallizing system during a loading-unloading cycle (Fig. 34a), the dichroic functions $(R - 1)/(R + 2)$ [66] of the 1126 cm^{-1} and 1662 cm^{-1} bands, respectively, have been plotted versus strain in Fig. 39. The representation of the spectroscopic data in this form has been chosen because no values for the transition moment directions of these absorption bands were available from the literature. Nevertheless, the onset of strain-induced crystallization and its retention during unloading can be readily derived from the dichroic function plot of the 1126 cm^{-1} absorption band. Samuels [66] has shown in detail, that if the dichroic functions for two different bands (a, b) from the same sample are plotted against each other, the slope of the resultant line will depend on the relation between their respective orienting phases and transition moment angles, since

$$\frac{[(R - 1)/(R + 2)]_a}{[(R - 1)/(R + 2)]_b} = \frac{f_a}{f_b} \cdot \frac{[(R_0 + 2)/(R_0 - 1)]_b}{[(R_0 + 2)/(R_0 - 1)]_a} \tag{25}$$

Figure 40 shows the computer-plots of Eq. (25) in terms of the 1126 cm^{-1} and 1662 cm^{-1} absorption bands for the elongation half-cycles of two independent

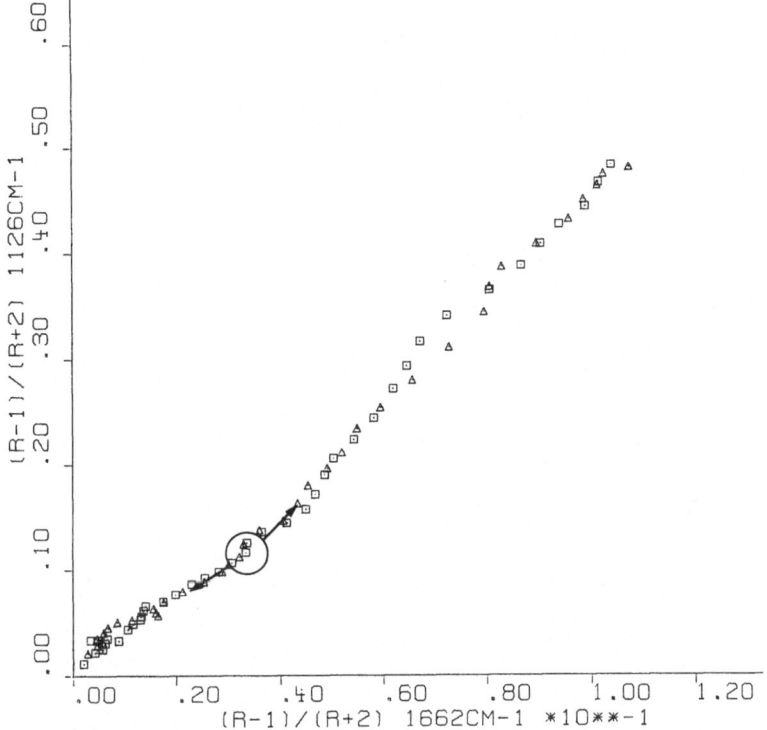

Fig. 40. Computer-plot of the dichroic functions of the 1126 cm^{-1} and 1662 cm^{-1} absorption bands monitored during two independent elongation half-cycles ($\square$/$\triangle$) of sulfur-crosslinked natural rubber at 300 K

experiments at 300 K. This curve can be roughly decomposed into two straight lines intersecting at the point which is characteristic of the onset of strain-induced crystallization. While the straight line left of the intersection corresponds to the strain interval where the contributions of both absorption bands can be assigned to the amorphous phase (0–230 % strain), the steeper slope to the right of the intersection is a consequence of the newly developing strain-induced crystal phase. Thus, from the smaller slope of the straight line corresponding to the amorphous phase and under the assumption that the transition moment angle ψ of the 1126 cm^{-1} band is zero (the dichroic function is then equivalent to the orientation function of this absorption band) an angle of approximately 40° was derived for the transition moment direction of the $\nu(C=C)$ absorption band at 1662 cm^{-1} and the polymer chain axis. The drastic improvement of chain alignment in the strain-crystallizing regions relative to the amorphous polymer is reflected by the increase of the slope to the right of the inflection point by a factor of about 1.6. From the orientation and dichroic functions of the 1126 cm^{-1} and 1662 cm^{-1} bands, respectively, an average inclination angle of approximately 35° was calculated for the polymer chains of the crystal phase and the direction of stretch whereas a value of about 44° was determined for the chain alignment of the amorphous polymer in the 530 % drawn sample.

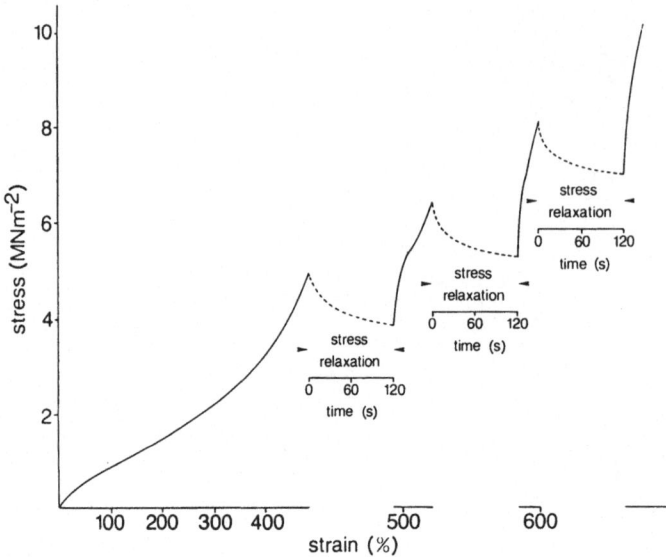

Fig. 41. Stress-strain curve of an elongation/stress-relaxation sequence of sulfur-crosslinked natural rubber at 300 K

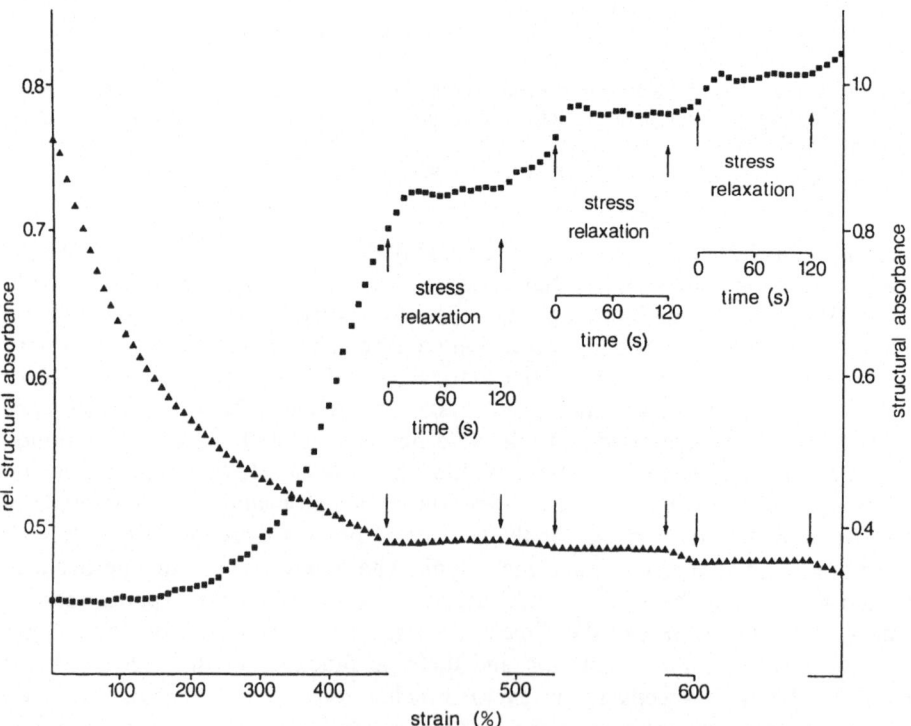

Fig. 42. $A_{0_{1662}}$ (▲) and $A_{0_{1126}}/A_{0_{1662}}$ ratio (■) monitored as a function of strain and stress-relaxation during the mechanical treatment shown in Fig. 41

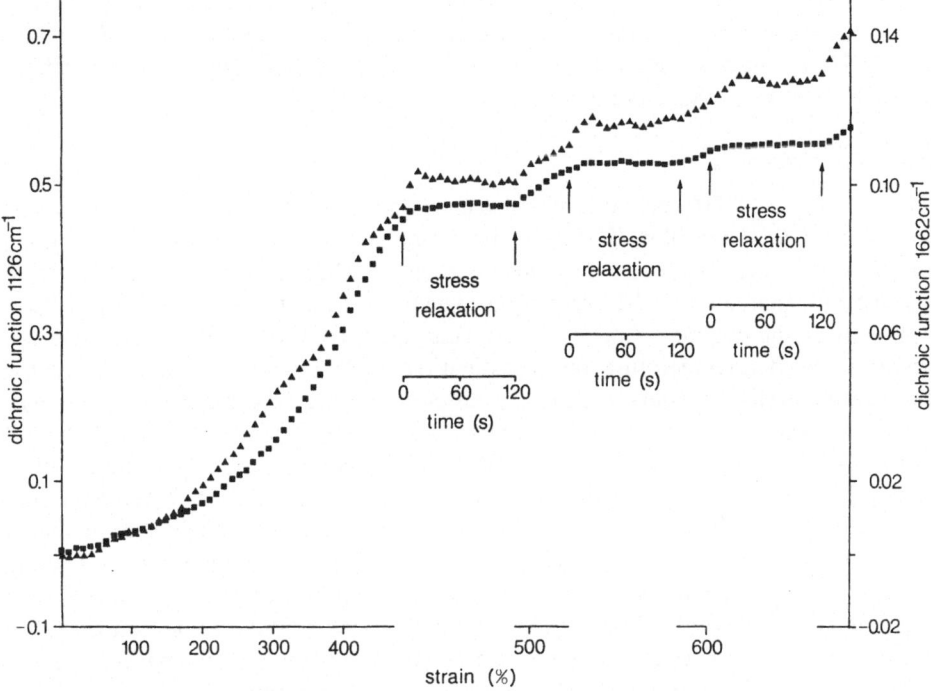

Fig. 43. Dichroic functions of the $1126\ cm^{-1}$ (■) and $1662\ cm^{-1}$ (▲) absorption bands monitored during the elongation/stress-relaxation sequence of Fig. 41

In separate experiments the influence of stress-relaxation on strain-induced crystallization and chain orientation was investigated by monitoring the $1126\ cm^{-1}$ and $1662\ cm^{-1}$ absorption bands as a function of relaxation time at various stress levels during elongation of a sulfur-crosslinked natural rubber film sample up to fracture. The stress-strain diagram of such an experiment is shown in Fig. 41. The corresponding structural absorbance ratios $A_{0_{1126}}/A_{0_{1662}}$ and the structural absorbance values $A_{0_{1662}}$ evaluated from the spectra taken simultaneously to the mechanical treatment are illustrated in Fig. 42 and reveal an interesting result. Thus, despite constant elongation the extent of crystallization further increases during the fast decay region of the stress-relaxation interval before it attains a maximum value. While an analogous behaviour can be observed for the dichroic function of the $1662\ cm^{-1}$ absorption band a distinctly smaller relative effect was obtained when the dichroic function of the $1126\ cm^{-1}$ crystallinity band is plotted as a function of stress-relaxation time (Fig. 43). From these results it can be concluded that in the crystalline phase the molecular orientation is completed almost immediately after elongation, while in the amorphous phase the polymer chains are further oriented during the fast decay region of stress-relaxation. These findings are in good agreement with previously reported results of rheo-optical IR orientation measurements of natural rubber [203] and suggest that the stress-relaxation can be primarily ascribed to the amorphous rather than the crystalline orientation.

Rheo-optical FTIR spectroscopy, therefore, not only provides a means to monitor strain-induced crystallization on-line to the mechanical treatment but also yields detailed information in terms of the orientation of the polymer chains in the amorphous polymer relative to those in the strain-crystallizing domains.

7.2.2 Polyurethanes

The rheo-optical FTIR technique has also been applied to study segmental orientation in polyester urethane films during uniaxial elongation and recovery. In fact, polyurethanes are particularly suited to such investigations because they contain functional groups with characteristic IR absorptions which can be assigned to the hard and soft segments of the polymer. Additional information becomes available from measurements at elevated temperature and investigations of NH-deuterated samples because the isotope exchange offers a means to differentiate the hard segments into phase-separated species and moieties dispersed in the soft segments (see chapter 5).

Linear polyurethanes are generally prepared by condensation of a diisocyanate with a high-molecular-weight macroglycol (commonly a polyether or polyester with a molecular weight between 1000 and 2000) and a low-molecular-weight chain extender dialcohol or diamine. By an appropriate choice of the type and relative amounts of these basic components the chemical and physical properties of the product can be regulated in a wide range. The following results have been obtained on samples taken from a series of polyester urethanes synthesized from diphenylmethane-4,4'-diisocyanate, a dihydroxy-terminated adipic acid/butane diol/ethylene glycol polyester (MW 2000) and butane diol with polyester:chain extender:diisocyanate molar ratios of 1.0:2.2:3.4 (a), 1.0:5.4:6.6 (b) and 1.0:7.5:8.7 (c), respectively. The soft segments of this type of polyurethane basically consist of the reaction products of the diisocyanate component and the macroglycol, whereas the hard segments contain largely aromatic and butane diol moieties linked together by urethane groups.

The segmental structure of polyurethanes becomes apparent from Fig. 44 which shows the linear polymer primary chains made up of alternating hard and soft segments. However, in the solid polymer the polymer primary chains do not really exist separately but rather the hard segments strongly tend to associate with each other through hydrogen bonding and aromatic π-electron attraction. As a consequence, the hard segments form domains in the mobile soft-segment matrix and a two-phase system results. The separate hard-segment domains crosslink the linear polymer

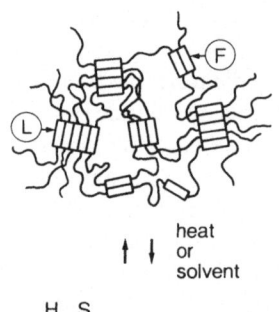

heat
or
solvent

H   S

Fig. 44. Schematic drawing of phase separation in a polyurethane elastomer (H: hard segments, S: soft segments, F: fibrillar hard segments, L: lamellar hard segments)

primary chains in lateral direction and extend the chains in linear fashion. The combination of these lateral and linear effects produces a crosslinked network which accounts for the elastic character of the polymer. These virtual crosslinks can be reversibly overcome by heat or solvation whereupon the polymer primary chains are more or less regenerated. Generally, phase segregation is a time-dependent phenomenon whose consequences cannot be overlooked in the characterization and use of these polymers. Numerous studies are available on the segmental structure of polyurethanes [70, 87, 95, 100, 210–244] particularly the hard-segment domains which might be expected to be crystalline but do not appear to be by conventional crystallinity tests such as wide-angle X-ray diffraction. It seems that the strong mutual attraction of the hard segments restricts their ability to organize themselves into a crystalline lattice. In this respect primarily small-angle X-ray scattering [210, 213, 219, 222, 228, 237, 240] and electron microscopy [228, 234–236] have contributed to a better knowledge of the existence, size, order, and separation of these domains.

Here, the results of rheo-optical investigations of the three abovementioned polyester urethanes in combination with elevated temperature measurements and deuterium exchange will be discussed [74, 245].

Film samples of the different polyester urethanes were prepared under identical conditions with a thickness of approximately 0.010 mm by casting from 2% w/v dimethylformamide solutions on surface-roughened glass plates and drying at 323 K in vaccum for 6 h. The film samples were then peeled off the glass plates in hot water, treated with boiling water for 0.5 h (for the removal of solvent traces) and finally dried in vacuum at 323 K for 0.5 h. The densities of the different polyester urethanes as measured in a gradient column (carbontetrachloride/chlorobenzene) were 1.236 g cm^{-3} (a), 1.251 g cm^{-3} (b) and 1.257 g cm^{-3} (c), respectively.

In the mechanical treatment film specimens of 12 mm length and 10 mm width were elongated at a draw rate of 72% strain/minute to 220% strain and then unloaded at the same rate to zero stress and 12-scan spectra were taken in 14-second intervals

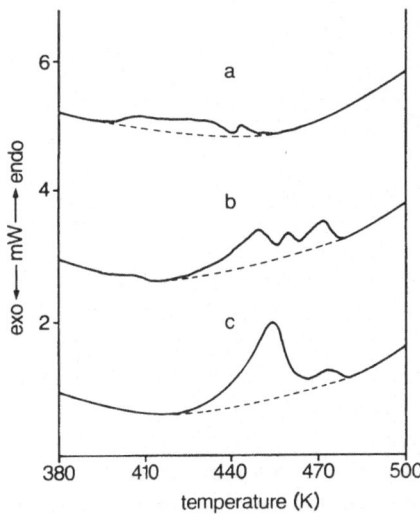

Fig. 45a–c. DSC diagrams of the investigated polyester urethane films based on different polyester: chain extender: diisocyanate molar ratios: **a** 1.0:2.2: 3.4; **b** 1.0:5.4:6.6; **c** 1.0:7.5:8.7

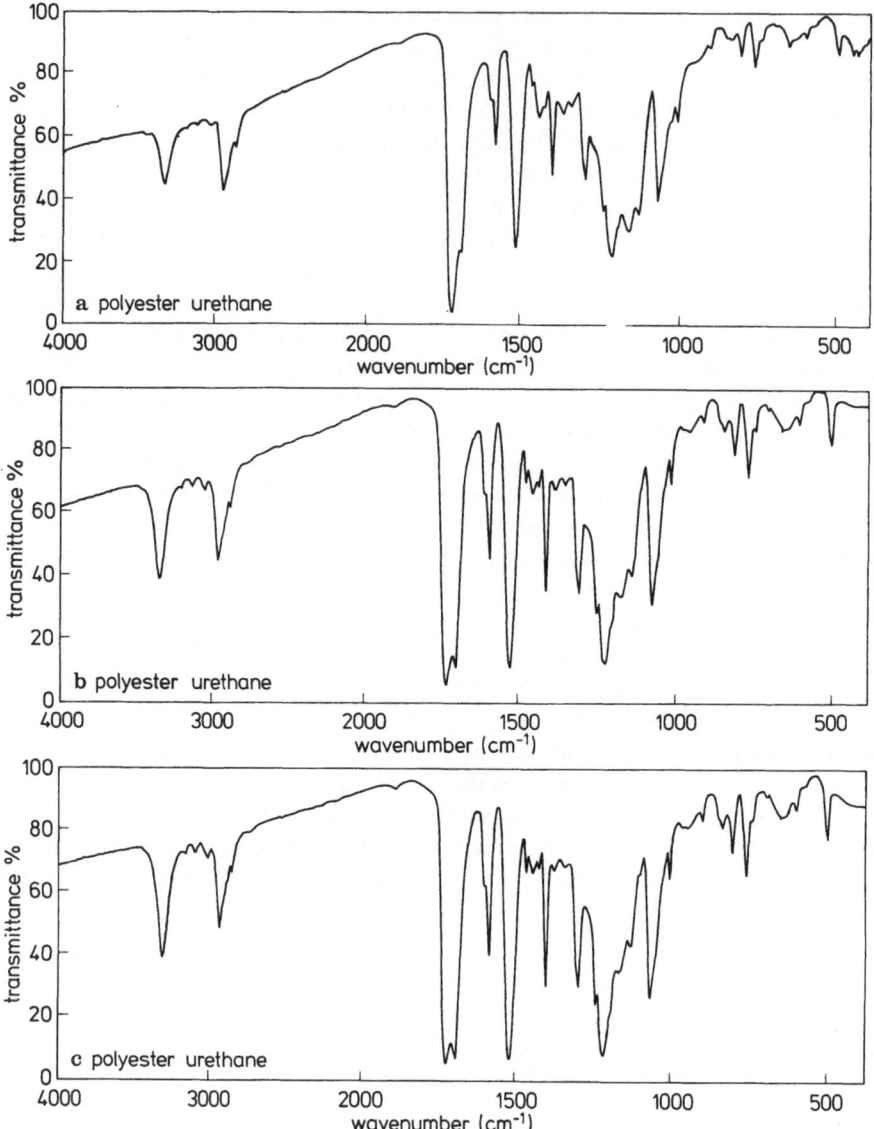

Fig. 46a–c. IR spectra of the investigated polyester urethanes **a**, **b**, and **c**

at a resolution of $2\,\mathrm{cm}^{-1}$. The rheo-optical experiments were performed at 300 K and 348 K for the original samples and at 300 K for the deuterated polymers. The deuterated specimens were prepared in the deuteration cell (Chapter 5) over a period of five days and finally transferred to the sample chamber of the stretching machine.

The DSC diagrams of the three polyester urethanes recorded between 380 K and 500 K with a heating rate of 10 K/min are shown in Fig. 45. The observed endotherms with multiple peaks reflect the sequential melting and disruption of the hard segment domains with different degrees of structural organization. Within the composition range studied, increasing the hard segment content resulted in higher ΔH values of 4.8 Jg^{-1} (a), 7.9 Jg^{-1} (b) and 13.7 Jg^{-1} (c). These ΔH values represent only about 15 % (a), 17 % (b) and 25 % (c), respectively, of that value expected from a model hard segment [246] and demonstrate the significantly reduced crystallization tendency of the hard segments of the investigated polyester urethanes.

In the wide angle X-ray diagrams only for the polyester urethane (c) with the largest hard segment content the Bragg reflection at about 0.75 nm [219, 237] could be detected apart from the intense amorphous halo at about 0.45 nm. Furthermore, no strain-induced crystallinity of the soft segments was observed in the 200 % elongated samples.

The IR spectra of the investigated polyester urethanes are shown in Fig. 46. On the basis of the established frequency correlations for the functional groups of poly-urethanes [70, 72] the extent of soft and hard segment orientation can be monitored by means of the polarization properties of the $\nu(CH_2)$ (2959 cm^{-1}), $\nu(C=O)_{ester}$ (1733 cm^{-1}) and $\nu(NH)$ (3331 cm^{-1}), $\nu(C=O)_{urethane}$ (1703 cm^{-1}), and $\delta(NH)$ + + $\nu(CN)$ (1531 cm^{-1}) absorption bands, respectively. The analysis of the $\nu(C=O)$ absorption region, however, is complicated by the overlap of the hydrogenbonded ester- and urethane-carbonyl absorptions [70, 72]. Owing to their almost uncoupled nature and separate frequency positions only the $\nu(NH)$ and $\nu(CH_2)$ absorption bands have been utilized for the quantitative evaluation of segmental orientation in the present study. The transition moment angles for both vibrations have been taken as 90° although it is recognized that some deviation of this value may occur for the $\nu(CH_2)$ stretching vibration due to superposition with the corresponding wagging mode. Under these assumptions the orientation functions for the hard and soft segments are represented by Eq. 11 b. In view of the hard segment crystal structure propos-ed by Blackwell [237], however, the maximum value to be expected for the $\nu(NH)$ orien-tation function of the hard segments is only about 0.65. Furthermore it should be kept in mind that the intensity contributions of the $\nu(CH_2)$ absorption bands of the chain extender and the diphenylmethane functionality increasingly contribute to the $\nu(CH_2)$ absorption intensity of the soft segments in the sequence polyester urethane (a) to (c). Nevertheless, the principal differences of hard and soft segment orientation can be readily derived from the spectroscopic data.

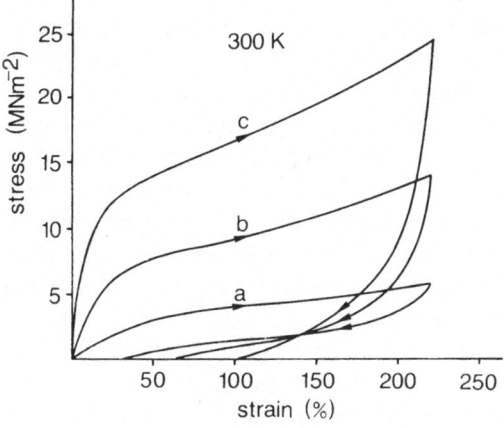

Fig. 47. Stress-strain curves of the loading-unloading cycles of the polyester urethane films with different hard and soft segment composition measured at 300 K

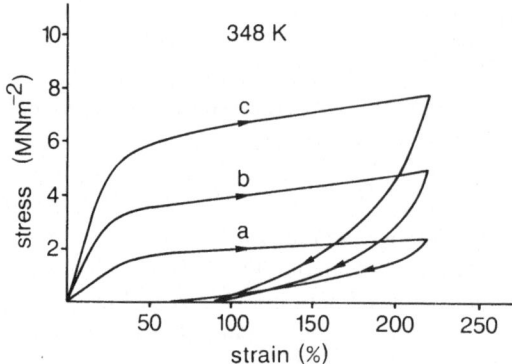

Fig. 48a–c. Stress-strain diagrams of the loading-unloading cycles of the polyester urethanes **a**, **b** and **c** measured at 348 K

The close relation between the composition and the mechanical properties of these polymers is reflected in the stress-strain diagrams measured at 300 K and 348 K (Figs. 47 and 48). Hence, at ambient temperature for the specified experimental conditions a distinct increase of initial modulus (11, 45 and 120 MNm^{-2}), stress-hysteresis (ratio of area bounded by a strain cycle to the total area underneath the elongation curve: 60, 80 and 90 %) and extension set (30, 65 and 100 %) can be observed with increasing hard segment content of polyester urethane (a) to (c).

The consequence of temperature elevation is a drastic decrease of stress level and initial modulus (5 MNm^{-2} (a), 18 MNm^{-2} (b), and 24 MNm^{-2} (c)). Furthermore, an increase of extension set (65 % (a) and 85 % (b)) and an increase and constancy, respectively, of stress hysteresis (68 % (a) and 80 % (b)) have been evaluated for the low and intermediate hard segment polymers (a) and (b) while a decrease of these parameters was observed for polyester urethane (c) (extension set: 85 %, stress hysteresis: 80 %) in comparison to the measurements at ambient temperature. Thus, at 348 K extension set and stress hysteresis reach their maximum values already for polyester urethane (b) with intermediate hard segment content [247].

The stress-strain diagrams of the deuterated specimens taken at 300 K did not deviate significantly from the mechanical data of the undeuterated polymers and are not shown separately.

Recently, Bonart et al. [248] have shown that the extent of orientation of the hard segment phase during elongation depends on the morphology of the hard segment domains and the interrelationship of two deformation mechanisms based on a morphological and a molecular level. As long as the chains of the soft segments are randomly coiled the matrix can be regarded as a continuum and the hard segment domains will be oriented by a continuum mechanical transfer of stress with their long axis dimension into the direction of stretch. Therefore, small, fibrillar hard segments in which the long axis dimension coincides with the polymer chain axes (see Fig. 44, (F)) will take up a positive orientation while lamellar domains with their long axis dimension perpendicular to the polymer chain axes (see Fig. 44, (L)) will be negatively oriented. This deformation mechanism dominates at low strains up to about 150 % strain and the strain value of the maximum negative orientation depends on the stability of the lamellar morphology and the length of the soft segments. With increasing extension of the soft segments a molecular transfer of stress by indi-

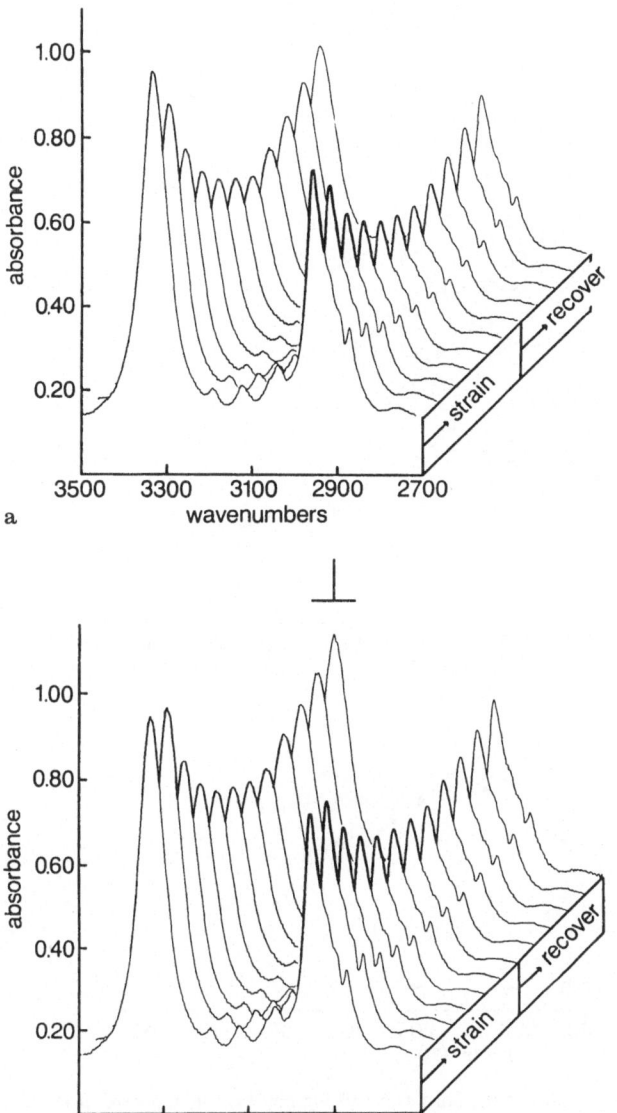

Fig. 49a and b. FTIR spectra of polyester urethane (c) film in the $\nu(NH)$ and $\nu(CH_2)$ stretching vibration region recorded at 300 K during uniaxial elongation to 220% strain and subsequent recovery to zero stress with radiation polarized alternately parallel and perpendicular to the direction of elongation

vidual chains becomes operative. For lamellar hard segment domains this mechanism leads to a disruption of the initially transversely oriented structural units with a subsequent positive orientation of the fragments into the direction of stretch. In the case of fibrillar hard segments continuum and molecular mechanical stress transfer both contribute to a positive alignment of the polymer chains.

To illustrate the dichroic effects the polarization spectra taken during a loading-unloading cycle of polyester urethane (c) at 300 K in the 3500–2700 cm^{-1} wavenumber region are shown in a stacked plot in Fig. 49. In Fig. 50 the corresponding orientation functions of the $\nu(NH)$ and $\nu(CH_2)$ absorption bands have been plotted for the

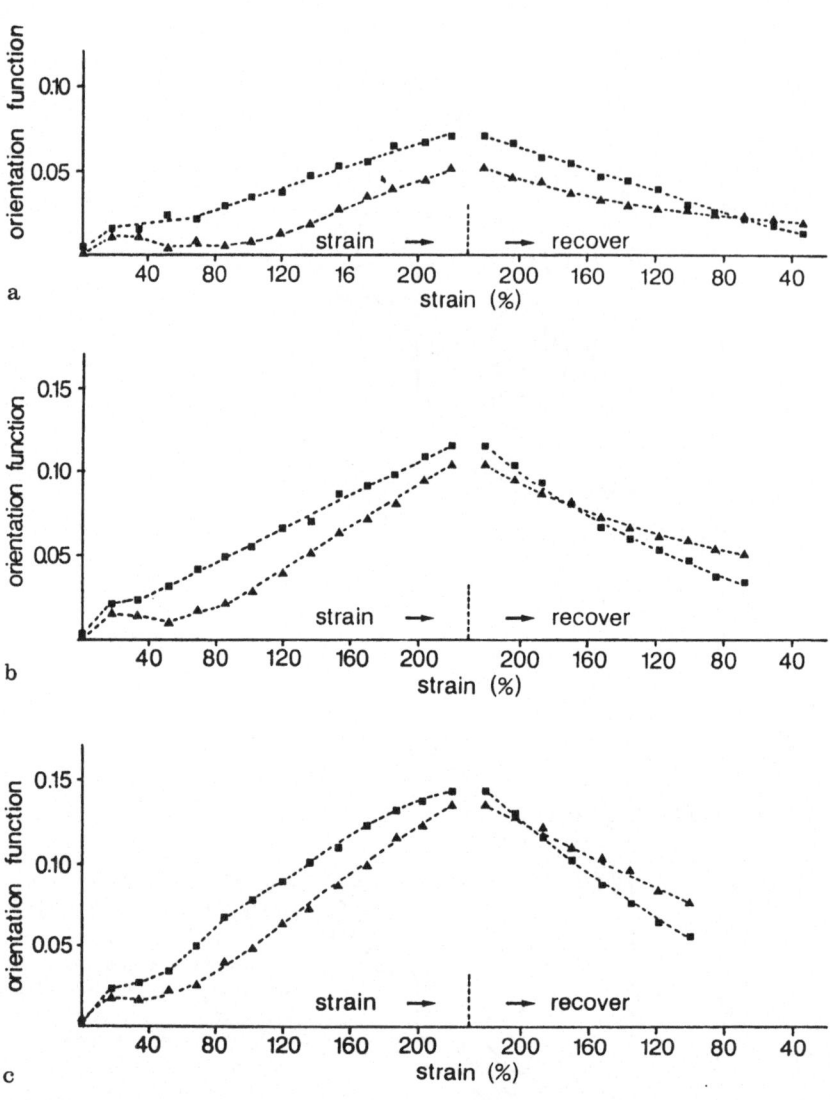

Fig. 50a–c. Orientation function/strain-plot of the hard and soft segments of the polyester urethanes a–c as monitored by the v(NH) (▲) and v(CH$_2$) (■) absorption bands, respectively, at 300 K

investigated polyester urethanes versus strain. Despite the comparatively low values of these orientation functions (see above) distinct differences between the hard and soft segment orientation, respectively, can be derived both during elongation as well as recovery with increasing hard segment content. Generally, up to the maximum elongation of 220% strain the soft segments exhibit a better average alignment in the direction of stretch than the hard segments. Additionally, the soft segments directly respond to the application of stress with an almost linear increase in positive chain orientation as a function of strain. In contrast, an initial strain interval with orientation function values in the vicinity of zero is observed for the hard segments

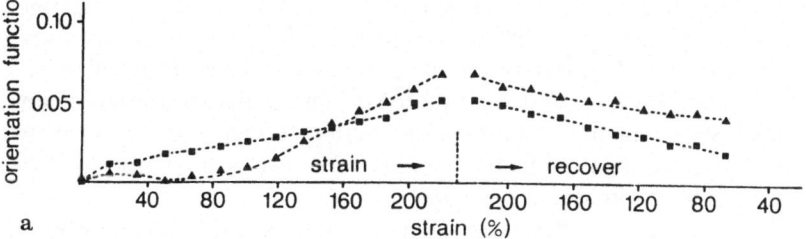

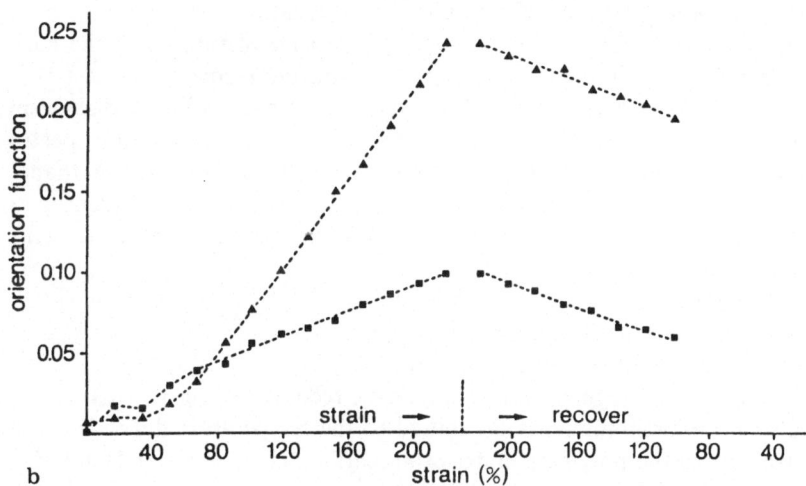

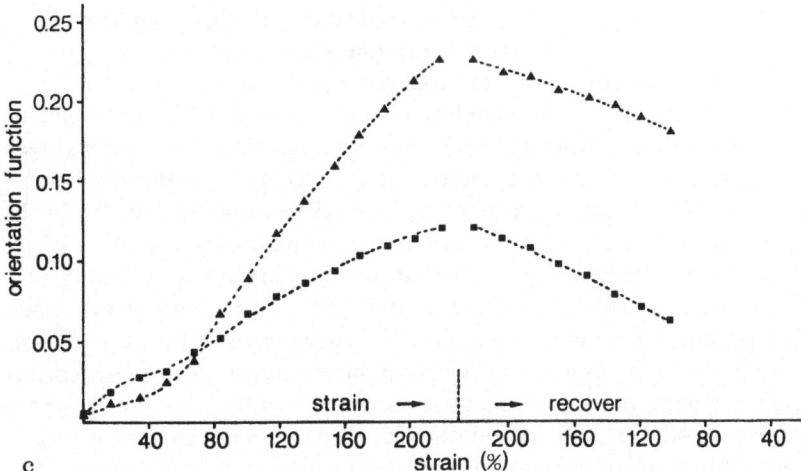

Fig. 51a–c. Orientation function/strain-plot of the hard and soft segments of the polyester urethanes **a–c** as monitored by the ν(NH) (▲) and ν(CH₂) (■) absorption bands, respectively, at 348 K

before the onset of significant positive orientation [249]. This difference can be explained in terms of the abovementioned antagonism of lamellar hard segment alignment during elongation. In this region obviously the positive orientation of small fibrillar hard segments is compensated by the negative orientation of lamellar hard segment domains. The shift in the onset of significant positive orientation to higher strain values for increasing soft segment content (polyester urethane (c) to (a)) is an indication of the corresponding increase in soft segment length.

Upon recovery to zero stress the orientation is more effectively retained by the hard segments. This phenomenon may be attributed to the entropy-driven relaxation and flow of the soft chain segments during unloading. As the soft segments relax they exert a tension on the hard segments thereby imposing an additional barrier to their recovery. The larger amount of this residual orientation for increasing hard segment proportion is the consequence of a more extensive disruption of the hard segments during elongation and their reorganization during recovery.

The orientation functions of the hard and soft segments derived from the dichroism of the ν(NH) and ν(CH$_2$) absorption bands, respectively, in the polarization spectra monitored during the loading-unloading cycles of the different polyester urethanes at 348 K are shown in Fig. 51. Basically, the following structural consequences of the mechanical treatment at 348 K are observed with respect to the measurements at ambient temperature:

a) earlier onset of positive hard segment orientation
b) drastic enhancement of hard segment orientation
c) slight deterioration of soft segment orientation
d) larger retention of hard segment alignment upon recovery to zero stress.

These effects can be predominantly attributed to the temperature dependence of the domain structure in the polymers under examination. Thus, it was indicated [223] that upon heating polyurethanes above about 343 K the degree of domain formation gradually decreases such that more mixing occurs between hard and soft segments. The abovementioned weakening of the hydrogen bonds between 300 K and 348 K will also contribute to an increased disruption tendency of the hard segments upon application of stress. While this enhanced disruption tendency of the hard segments leads to an earlier onset of their more pronounced positive orientation at elevated temperature on the other hand a somewhat lower chain alignment of the soft segments is effected during elongation. Owing to the low crystallization tendency of the involved soft segments strain-induced crystallization does not account for the mechanical behaviour of polyester urethane (c). Therefore, it could be assumed that due to the higher state of order in the hard segments of polyester urethane (c) (see also Fig. 45) the smaller domains are preferentially disrupted during elongation at elevated temperature [239] resulting in a smaller positive hard segment orientation at maximum elongation than polyester urethane (b). The smaller extension set and stress hysteresis of polyester urethane (c) in comparison to the ambient temperature measurements may be interpreted in terms of a more homogeneous stress distribution in the elongated sample and the improved retractive force of the soft segments at elevated temperature.

The orientation functions of the hard and soft segments evaluated from the polarization spectra taken during the loading-unloading cycles of the deuterated polyester urethanes at 300 K are shown in Fig. 52. Here, the ν(NH) orientation function represents the orientation of the inaccessible hard segment domains. According to Bonart's

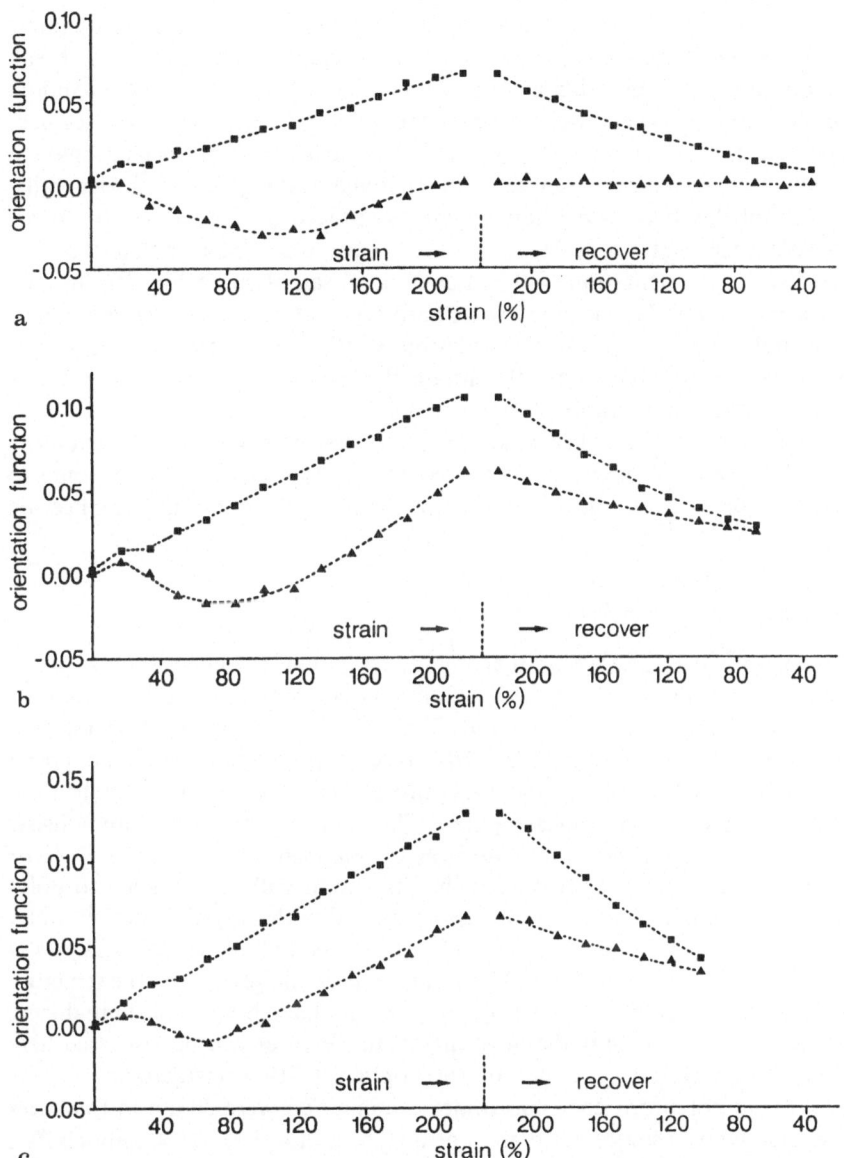

Fig. 52a–c. Orientation function/strain-plot of the hard and soft segments of the deuterated polyester urethanes **a–c** as monitored by the $\nu(NH)$ (▲) and $\nu(CH_2)$ (■) absorption bands, respectively, at 300 K

model of segmental orientation in polyurethanes [248] the pronounced transverse hard segment orientation observable in Fig. 52 is therefore fully consistent with the assignment of the involved urethane groups to phase-separated lamellar domains. As can be readily derived from Fig. 52 the negative orientation of the hard segments does not immediately start with the application of stress. This supports a structural model proposed by Cooper et al. [225a] in which the hard segment domains are partly inter-connected. Thus, the interconnections of such a hard segment network have to break up first before any transverse alignment can take place. The shift of the maximum negative orientation to higher strain values in the polymers with lower hard segment content [110% (a), 80% (b) and 70% (c)] is also indicative of the delayed onset of the molecular-mechanical stress-transfer owing to longer soft segments [239]. Generally, the orientation of the inaccessible hard segments induced during elongation to the maximum experimental strain of 220% has been found much smaller than the total average hard segment orientation of the undeuterated samples (see Fig. 50). In fact, in the polyester urethane (a) with the lowest hard segment content the lamellar hard segments do not take on a positive orientation at all. The deuteration technique, therefore, offers the possibility to differentiate the individual contributions to the average hard segment orientation.

In conclusion, rheo-optical FTIR studies of this class of polymers provide a more detailed insight into their orientation and recovery mechanisms and contribute to a better correlation and optimization of their chemical structure and engineering applications.

7.2.3 Hard-Elastic Polypropylene

In the mid-sixties it was discovered that highly crystalline polymers such as poly-propylene can be extruded under specific conditions of crystallization from the melt to produce remarkably elastic fibers and films [250-252]. Owing to their unusual mechanical properties they were termed hard-elastic or springy polymers. The typical processing features of hard-elastic materials involve melt-extrusion, crystallization under high stress and a final annealing step. The resulting superstructure consists of stacked lamellar aggregates whose surfaces are aligned normal to the fiber or film extrusion direction [113, 253]. Although the discussion will be restricted to poly-propylene here, a similar behaviour has been observed with polyethylene [254], poly-pivalolactone [255, 256], poly-4-methylpentene, poly-3-methylbutene [257] and poly-amide-6,6 when specially annealed [258]. Numerous papers and reviews on the structure and the mechanical properties of this class of polymers have been publishing during the last decade [259-266] and only the most important features will be discussed here with reference to recent data obtained by rheo-optical FTIR investigations.

The hard-elastic polypropylene films under examination were 30 μm in thickness and had been previously subjected to an annealing step at 413 K for 30 minutes [267]. The density of the samples as measured in the gradient column (methanol) was 0.899 g cm^{-3} and from DSC measurements a heat of melting of 93.8 Jg^{-1} was deter-mined. For the rheo-optical measurements film strips were cut 7 mm in width and 15 mm in length. To suppress the interference fringes in the IR spectra of the original specimens their surface had to be roughened slightly thereby reducing the thickness to about 25 μm. Two alternative mechanical treatments were applied:
a) elongation up to 130% strain
b) elongation to 70% strain and subsequent recovery to zero stress.

The elongation and recovery rate was 33% per minute and simultaneously 8-scan FTIR polarization spectra were taken in 6.5-second intervals with a resolution of 4 cm^{-1}.

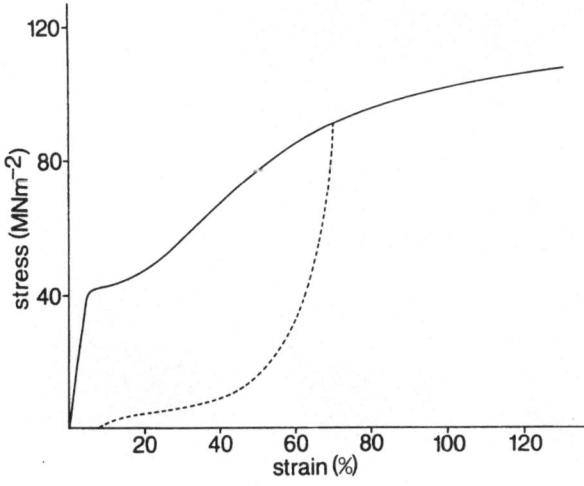

Fig. 53. Stress-strain diagram of hard-elastic polypropylene film measured at 300 K (——— elongation, — — — recoverey)

The stress-strain curves corresponding to the abovementioned mechanical treatments are shown in Fig. 53 where the dashed unloading trace reflects the excellent elastic recovery from high strains. Generally, the elongation of hard-elastic, annealed polypropylene is characterized by an initial steep, linear increase of stress, followed by a sharp yield point at about 5% strain. A second, diffuse yield point is observable in the vicinity of 60% strain. While the elastic recovery is of the order of 90% on the first cycle, the work recovery is relatively low at about 24% (Fig. 53). This is due to a large amount of energy required to overcome the van der Waals forces in the intercrystalline regions during the first 5–10% extension. On immediate and repeated cycling the stress-level of the yield point at about 5% strain is considerably lowered and the work recovery increases to about 60% within a few cycles [263]. Furthermore, these polymers are characterized by a negative temperature coefficient of retractive force on stretched materials and high deformability with good recovery at liquid nitrogen temperature [257, 259, 260]. These and other data have led to the suggestion that the elastic restoring forces in these materials are energy- rather than entropy-driven [259, 260].

X-ray-diffraction, electron microscopy, light-scattering, and density measurements [259, 260, 266] show that the elastic characteristics are based on the development of voids between the lamellae. At the onset of their formation which coincides with the sharp yield-point at 5% strain the long axes of these microvoids are oriented perpendicular to the deformation direction. As extension proceeds, vertical microvoids are generated by yielding of horizontal microvoids with subsequent extension in the deformation direction.

Figure 54 shows typical X-ray patterns [266] of hard-elastic, annealed polypropylene films in the undrawn and extended state. The effect of elongation of springy polypropylene samples on their original wide-angle X-ray patterns which reflect the good polymer chain orientation parallel to the direction of stretch is slight; essentially, some of the arcs are observed to undergo very small extensions in the azimuthal direction with increased stretching. This suggests a slight disorientation or rotation of the diffracting crystals [266]. In the small-angle X-ray photographs it is noted

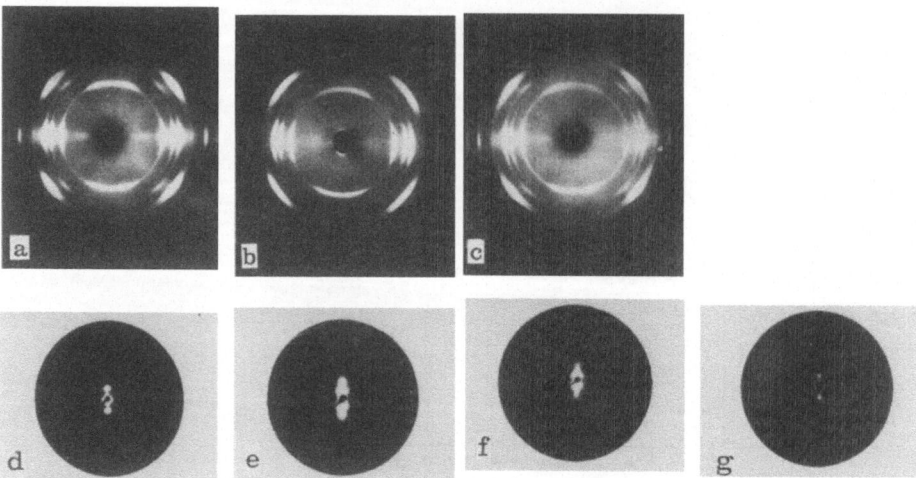

Fig. 54a–g. Wide-angle (**a–c**) and small-angle (**d–g**) X-ray diagrams of hard-elastic, annealed polypropylene; a–c: polypropylene film annealed for 0.75 h at 140 °C, **a** undrawn, **b** 73 % extension, **c** recovered after extension to 73 % strain; **d–g**: polypropylene fiber annealed for 1 h at 150 °C, **d** undrawn, **e** 50 % extension, **f** 100 % extension, **g** recovered after extension to 100 % strain [reproduced from Noether, H. D., W. Whitney, Kolloid-Z. und Z. Polymere, *251*, 991–1005 (1973) by permission of the publishers, Dr. Dietrich Steinkopff Verlag, Darmstadt, West Germany)

that the scattering is along the meridian and consists of an intense and a weak pair of spots [260]. Whereas the intense diffraction increases and decreases its spacing on stretching and retraction and has been associated with the interlamellar distance [257], the less intense pair of spots maintains its spacing constant on extension and has been correlated with the lamellar thickness because of its dependence on annealing conditions [266].

The ability of springy polymers to maintain their high elastic recovery down to low temperatures has led to the formulation of a large number of energy models for their deformational behaviour. Most of these mechanisms such as for example the leaf spring model [260, 266], or the reversible shear of lamellae between fixed tie-points [262], depend upon the assumption of lamellar bending. Elastic, restorative action has been postulated due to such diverse mechanisms as van der Waals forces, fold repulsion, and increased surface energy [29]. However, Göritz and Müller [265] showed by simultaneous thermal and mechanical measurements that there is also evidence for entropic contributions to the retractive forces. Thus, heat absorption decreases after extension to about 5 % strain and internal energy begins to increase at the same extension level. The actual mechanism behind the unique behaviour of this class of polymers is undoubtedly a combination of the proposed models with entropic as well as energetic contributions [254].

The rheo-optical measurements have been primarily performed with the intention to obtain further experimental evidence to support or reject specific features of the existing models. The FTIR polarization spectra taken during elongation up to 130 % strain are shown in Fig. 55. The evaluation of these spectra in terms of the state of order and the orientation functions of the crystalline regions and the average polymer

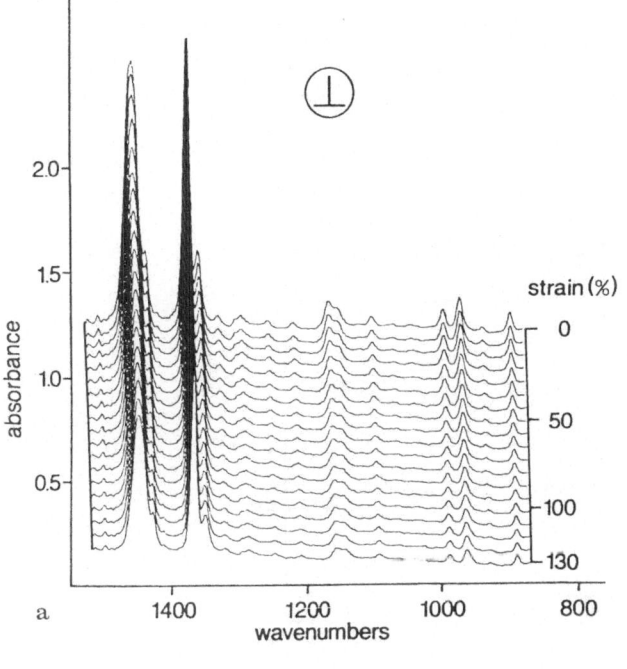

a

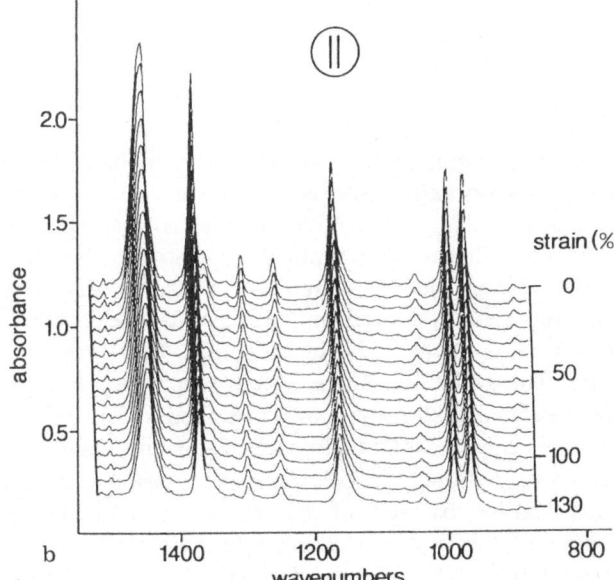

b

Fig. 55a and b. FTIR polarization spectra of a hard-elastic polypropylene film recorded during extension to 130 % strain with light polarized alternately parallel and perpendicular to the direction of elongation

are shown in Fig. 56 and 57, respectively. In Fig. 56 the structural absorbances of the 998 cm^{-1} and 972 cm^{-1} bands are shown alongside the corresponding ratio $A_{0_{998}}/A_{0_{972}}$ as a function of strain. As has been pointed out in section 7.1.2 already the 998 cm^{-1} band is representative of conformationally regular chains in the ordered

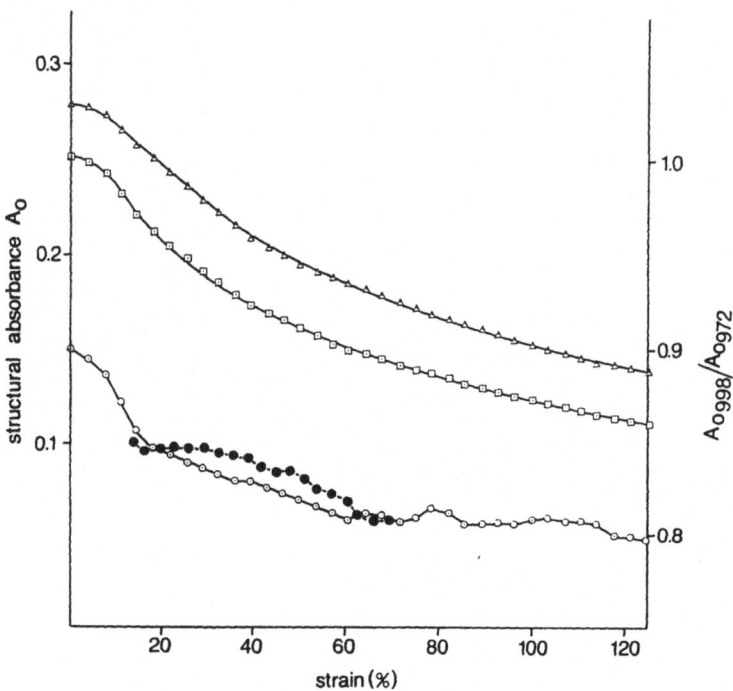

Fig. 56. Structural absorbances of the 998 cm^{-1} ($\square$) and 972 cm^{-1} ($\triangle$) absorption bands and the corresponding ratio $A_{0_{998}}/A_{0_{972}}$ ($\bigcirc$) as a function of strain (open symbols: elongation, closed symbols: recovery)

domains while the 972 cm^{-1} absorption contains contributions of both phases. Thus, the initial decrease of the structural absorbance ratio is — in good agreement with the observations of Göritz and Müller [265] — indicative of a partial melting of — predominantly interlamellar — crystallites. Upon immediate recovery, however, only a partial recrystallization is observed in the sample under examination. This discrepancy to the reversibility reported by Göritz and Müller may be primarily attributed to the fact that in the rheo-optical FTIR investigations the sample was extended to much higher strains. The evaluation of the spectroscopic data in terms of the orientation functions of these two absorption bands (see Section 7.1.2) shows that neither in the crystalline nor in the amorphous domains significant changes of orientation can be detected up to 40 % strain (Fig. 57). At that extension a gradual improvement of chain alignment can be observed for the absorption band which is characteristic of the average polymer while a comparable effect for the crystalline regions does not occur before 50 % strain. The principle difference between the small orientation phenomena in the amorphous and crystalline phases, however, is the reversibility observed for the amorphous domains during immediate recovery from 70 % strain (Fig. 53). It has been shown by small-angle X-ray investigations that the deformation range between 5–10 % and about 40–60 % strain is dominated by microvoid formation accompanied by the separation of the lamellar aggregates [260, 261, 266]. In the course of this process elongation leads to a slight improvement of the

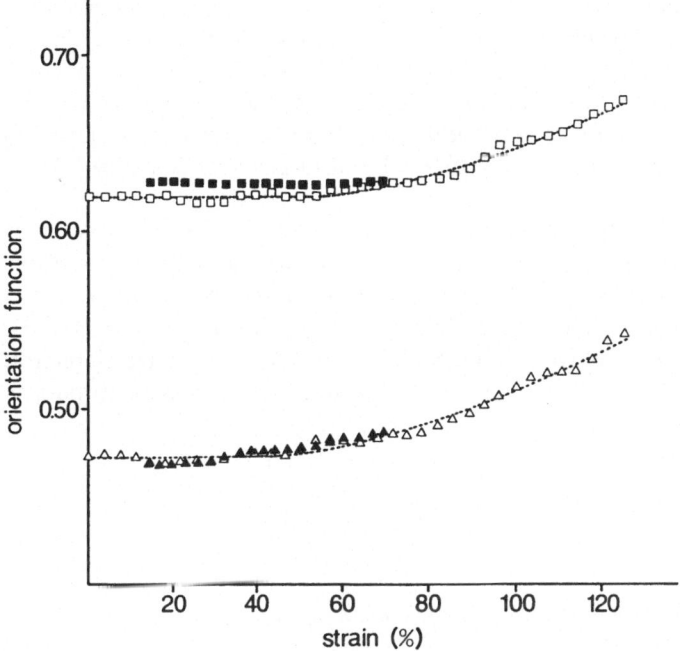

Fig. 57. Orientation function/strain-plot of the 998 cm^{-1} (□) and 972 cm^{-1} (△) absorption bands as a function of strain (open symbols: elongation, closed symbols: recovery)

orientation of the amorphous regions which is detectable in the FTIR polarization spectra. In the subsequent region of plastic deformation beyond the diffuse yield point at about 60 % strain individual chains or bundles are pulled out of the lamellae and may contribute to the irreversible orientation effects of conformationally regular domains.

8 Conclusions and Outlook

Originally applied predominantly as an analytical tool in the field of polymer characterization infrared spectroscopy has been increasingly utilized in the last decades for the elucidation of the physical structure of polymers. However, with the advent of rapid-scanning FTIR instruments and the development of the rheo-optical FTIR technique infrared spectroscopy has been launched into a completely new era of polymer physical applications.

Advances in rheo-optical FTIR spectroscopy and in the interpretation of the experimental results derived thereof are to be expected in the near future primarily in two aspects. First, the time scale for the acquisition of single-scan spectra with commercial instruments will be further pushed down. This improvement in time resolution automatically requires the replacement of a mechanically operated polarizer unit by a photoelastically modulated system [19, 108, 268–270] which more fully exploits the scanning speed of such an instrument by a much faster change of polarization direction between the individual scans. On the other hand, a considerable impact on the interpretation of the data obtained by rheo-optical FTIR spectroscopy is to be expected from a further progress in the development of dynamic X-ray investigations of polymer systems with synchrotron radiation. The combination of these fast-scanning techniques and the correlation of their results will certainly provide new aspects for the detection of structural changes in polymers during thermal and mechanical processes.

Acknowledgement: The author gratefully acknowledges the experimental assistance of H. Devrient, E. Knoll, H. P. Schlemmer and W. Schmitt and helpful discussions and the supply of polymer samples by Dr. G. Bayer, Prof. Dr. R. Bonart, Dr. U. Eisele, Dr. B. Heise, Dr. H. Hespe, Dr. W. Hoffmann, Dr. K. Holland-Moritz, Dr. H. Krömer, Prof. Dr. H. D. Noether, Dr. G. Spilgies and Dr. F. Schultze-Gebhardt. The author would also like to thank Bayer AG for granting permission to publish the experimental data.

9 References

1. Onogi, S., Asada, T.: Progr. Polym. Sci. Jap. (Imoto, M., Onogi, S. eds.), *2*, p. 261, New York: John Wiley 1971
2. Le Grand, D. G., Erhardt, P. F.: Trans. Soc. Rheol. *6*, 301 (1962)
3. Read, B. E.: Polymer *3*, 143 (1962)
4. Stein, R. S.: J. Polym. Sci. Part C, *15*, 185 (1966)
5. Takenaka, T., Shimura, Y., Gotoh, R.: Kolloid-Z. u. Z. Polymere *237*, 193 (1970)
6. Siesler, H. W., Holland-Moritz, K.: Infrared and Raman Spectroscopy of Polymers, New York: Marcel Dekker 1980
7. Siesler, H. W.: Proceedings of the 5th European Symposium on Polymer Spectroscopy (Hummel, D. O. ed.), p. 137, Weinheim: Verlag Chemie 1979
8. Holland-Moritz, K., Holland-Moritz, I., van Werden, K.: Colloid & Polym. Sci. *259*, 156 (1981)
9. Siesler, H. W.: J. Mol. Struct. *59*, 15 (1980)
10. Siesler, H. W.: Polym. Prep. Amer. Chem. Soc. Div. Polym. Chem. *21* (*1*), 163 (1980)
11. Siesler, H. W.: Proceedings of IUPAC 28th Macromolecular Symposium, University of Massachusetts, Amherst, USA, p. 46 (1982)
12. Zachmann, H. G.: ibidem p. 651
13. Forgacs, P., Sheromov, M. A., Tolochko, B. P., Mezentsev, N. A., Pindurin, V. F.: J. Polym. Sci. Phys. Ed. *18*, 2155 (1980)
14. Koch, M. J. H., Bordas, J., Schöla, E., Broecker, H. Ch.: Polymer Bulletin *1*, 709 (1979)
15. Annual Report 1982, Hamburger Synchrotronstrahlungslabor HASYLAB am Deutschen Elektronen-Synchrotron DESY, p. 155–176
16. Griffiths, P. R.: Chemical Infrared Fourier Transform Spectroscopy, New York: Wiley-Interscience, 1975
17. Geick, R.: Fresenius Z. Anal. Chem. *288*, 1 (1977)
18. Koenig, J. L.: Adv. Polym. Sci. *54*. 87 (1983)
19. Burchell, D. J., Lasch, J. E., Farris, R. J., Hsu, S. L.: Polymer *23*, 965 (1982)
20. Lasch, J. E., Masaoka, T., Burchell, D. J., Hsu, S. L.: Polymer Bulletin *10*, 51 (1983)
21. Fateley, W. G., Koenig, J. L.: J. Polym. Sci. Lett. Ed. *20*, 445 (1982)
22. Sakai, H., Murphy, R. E.: Appl. Optics *17*, 1342 (1978)
23. Mantz, A. W.: Appl. Optics *17*, 1347 (1978)
24. Schlemmer, H. P., Siesler, H. W.: unpublished results
25. Holland-Moritz, K.: private communication
26. Gubanov, A. I., Kosobukin, V. A.: Mekh. Polym. *4*, 586 (1968)

27. Zhurkov, S. N., Vettegren, V. I., Korsukov, V. E., Novak, I. I.: Phys. Tverd. Tela *11*, 290 (1969)
28. Roylance, D. K., DeVries, K. L.: J. Polym. Sci. *B9*, 443 (1971)
29. Vettegren, V. I., Novak, I. I.: J. Polym. Sci. Polym. Phys. Ed. *11*, 2135 (1973)
30. Wool, R. P., Statton, W. O.: J. Polym. Sci. Polym. Phys. Ed. *12*, 1575 (1974)
31. Gubanov, A. I., Kosobukin, V. A.: Mekh. Polym. *1*, 33 (1975)
32. Wool, R. P.: J. Polym. Sci. Polym. Phys. Ed. *13*, 1795 (1973)
33. Evans, R. A., Hallam, H. E.: Polymer *17*, 839 (1976)
34. Mitra, V. K., Risen, W. M., Baughman, R. H.: J. Chem. Phys. *66*, 2731 (1977)
35. Mocherla, K. K. R., Statton, W. O.: J. Appl. Polym. Sci. Appl. Polym. Symp. *31*, 183 (1977)
36. Voroboyev, V. M., Razumovskaja, I. V., Vettegren, V. I.: Polymer *19*, 1267 (1978)
37. Wool, R. P.: J. Polym. Sci. Polym. Phys. Ed. *19*, 449 (1981)
38. Kausch, H. H.: Polymer Fracture, Heidelberg: Springer Verlag 1978
39. Sikka, S. S., Kausch, H. H.: Colloid & Polym. Sci. *257*, 1060 (1979)
40. Sikka, S. S.: Matl. Sci. & Eng. *41*, 265 (1979)
41. Sikka, S. S.: Polymer Bulletin *3*, 123 (1980)
42. Levy, R. L.: Amer. Chem. Soc. Div. Polym. Chem. Prepr. *21*, 263 (1980)
43. Levy, R. L.: Org. Coatings & Plastics Chem. *45*, 727 (1981)
44. Zhurkov, S. N., Korsukov, V. E.: J. Polym. Sci. Polym. Phys. Ed. *12*, 385 (1974)
45. DeVries, K. L., Roylance, D. K., Williams, M. L.: J. Polym. Sci. *A1*, *8*, 237 (1970)
46. Nagamura, T., Fukitani, K., Takayanagi, M.: J. Polym. Sci. Polym. Phys. Ed. *13*, 1515 (1975)
47. Ranby, B., Rabek, J. F.: ESR Spectroscopy in Polymer Research, Heidelberg: Springer Verlag 1977
48. DeVries, K. L.: J. Appl. Polym. Sci. Appl. Polym. Symp. *35*, 439 (1979)
49. Frank, O., Wendorff, J. H.: Colloid & Polym. Sci. *259*, 1047 (1981)
50. Tabb, D. L.: Ph. D. Thesis, Case Western Reserve University, Cleveland, Ohio, USA, 1974
51. Stoeckel, T. M., Blasius, J., Crist, B.: J. Polym. Sci. Polym. Phys. Ed. *16*, 485 (1978)
52. Fanconi, B. M., DeVries, K. L., Smith, R. H.: Polymer Suppl. *23*, 1027 (1982)
53. Dechant, J.: Ultrarotspektroskopische Untersuchungen an Polymeren, Berlin: Akademie Verlag 1972
54. Zbinden, R.: Infrared Spectroscopy of High Polymers, New York: Academic Press 1964
55. Jasse, B., Koenig, J. L.: J. Macromol. Sci. Rev. Macromol. Chem. *C17*, 61 (1979)
56. Read, B. E.: Structure and Properties of Oriented Polymers (Ward, I. M. ed.), p. 150, London: Applied Science 1975
57. Fraser, R. D. B.: J. Chem. Phys. *21*, 1511 (1953)
58. Samuels, R. J.: Structured Polymer Properties, New York: Wiley-Interscience 1974
59. Peterlin, A. (ed.): Plastic Deformation of Polymers, New York: Marcel Dekker 1971
60. Petermann, J., Kluge, W., Gleiter, H.: J. Polym. Sci. Polym. Phys. Ed. *17*, 1043 (1979)
61. Heise, B., Kilian, H. G., Wulff, W.: Progr. Colloid Polym. Sci. *67*, 143 (1980)
62. Juska, T., Harrison, I. R.: Polym. Eng. Revs. *2(1)*, 13 (1982)
63. Fraser, R. D. B.: J. Chem. Phys. *24*, 89 (1956)
64. Hermans, P. H.: Contributions to the Physics of Cellulosic Fibers, p. 138, Amsterdam: Elsevier 1946
65. Alexander, L. E.: X-ray Diffraction Methods in Polymer Science, New York: Interscience 1969
66. Samuels, R. J.: Makromol. Chem. Suppl., *4*, 241 (1981)
67. Samuels, R. J.: J. Polym. Sci. *A3*, 1741 (1965)
68. Liang, C. Y.: Polymer Characterization (Ke, B. E. ed.), p. 33, New York: Wiley-Interscience 1964
69. Holland-Moritz, K., Siesler, H. W.: Appl. Spectrosc. Revs. *11*, 1 (1976)
70. Srichatrapimuk, V. W., Cooper, S. L.: J. Macromol. Sci. Phys. *B15*, 267 (1978)
71. Cannon, C. G.: Spectrochim. Acta *16*, 302 (1960)
72. Ishihara, H., Kimura, I., Saito, K., Ono, H.: J. Macromol. Sci. Phys. *B10*, 591 (1974)
73. Garton, A., Phibbs, M. K.: Makromol. Chem. Rapid Commun. *3*, 569 (1982)
74. Siesler, H. W.: Polymer Bulletin, *9*, 557 (1983)
75. Alter, H., Hsaio, H. Y.: J. Polym. Sci. Polym. Letters Ed. *6*, 663 (1968)
76. Benedetti, E., Verganini, P., Andruzzi, F., Maganini, P. L.: Polymer Bulletin *2*, 241 (1980)

77. Zachmann, H. G., Stuart, H. A.: Makromol. Chem. *44/46*, 622 (1961)
78. Hannon, M. J., Koenig, J. L.: J. Polym. Sci. Part. *A-2, 7*, 1085 (1969)
79. Suchov, F. F., Iliceva, Z. F., Slochotova, N. A.: Vysokomol. Soed. *11*, 851 (1967)
80. Dahme, A., Dechant, J.: Acta Polymerica *33*, 490 (1982)
81. Stach, W.: Ph. D. Thesis, University of Cologne, West Germany, 1982
82. Ward, I. M., Wilding, M. A.: Polymer *18*, 327 (1977)
83. Paik Sung, C. S., Schneider, N. S.: Macromolecules *10*, 452 (1977)
84. Ogura, K., Sobue, H.: Polymer J. *3*, 153 (1972)
85. Nissan, A. H.: Macromolecules *9*, 840 (1976)
86. Nakayama, K., Ino, T., Matsubara, I.: J. Macromol. Sci. Chem. *A* 3, 1005 (1969)
87. Seymour, R. W., Cooper, S. L.: Macromolecules *6*, 48 (1973)
88. Murthy, A. S. N., Rao, C. N. R.: Appl. Spectroscopy. Revs. *2*, 69 (1968)
89. Schuster, P., Zundel, G., Sandorfy, C. (eds.): The Hydrogen Bond, New York: North Holland 1976
90. Pimentel, G. C., Sederholm, C. H.: J. Chem. Phys. *24*, 639 (1956)
91. Allen, L. C.: J. Amer. Chem. Soc. *97*, 6921 (1975)
92. Walton, A. G., Blackwell, J.: Biopolymers, New York: Academic Press 1973
93. Kinoshita, Y.: Makromol. Chem. *33*, 1 (1959)
94. Frigge, K., Dechant, J.: Faserforsch. Textiltechn. *26*, 547 (1975)
95. West, J. C., Cooper, S. L.: J. Polym. Sci. Polym. Symp. *60*, 127 (1977)
96. Bessler, E., Bier, G.: Makromol. Chem. *122*, 30 (1969)
97. Senich, G. A., MacKnight, W. J.: Macromolecules *13*, 106 (1980)
98. Schroeder, L. R., Cooper, S. L.: J. Appl. Phys. *47*, 4310 (1976)
99. Brunette, C. M., Hsu, S. L., MacKnight, W. J.: Macromolecules *15*, 71 (1982)
100. Born, L., Hespe, H., Crone, J., Wolf, K. H.: Colloid & Polym. Sci. *260*, 819 (1982)
101. Krimm, S.: Adv. Polym. Sci. *2*, 51 (1960)
102. Snyder, R. G., Schachtschneider, R. H.: Spectrochim. Acta *19*, 85 (1963)
103. Tasumi, M., Shimanouchi, T.: J. Chem. Phys. *43*, 1245 (1965)
104. Kaiser, R.: Kolloid-Z. und Z. Polymere *149*, 84 (1956)
105. Read, B. E., Stein, R. S.: Macromolecules *1*(2), 116 (1968)
106. Glenz, W., Peterlin, A.: J. Polym. Sci. Part A 2, *9*, 1191 (1971)
107. MacRae, M. A., Maddams, W. F.: J. Appl. Polym. Sci. *22*, 2761 (1978)
108. Noda, I., Dowrey, A. E., Marcott, C.: J. Polym. Sci. Polym. Lett. Ed. *21*, 99 (1983)
109. Wedgewood, A. R., Seferis, J. C.: Pure & Appl. Chem. *55*, 873 (1983)
110. Holmes, D. R., Palmer, R. P.: J. Polym. Sci. *31*, 345 (1958)
111. Aggarwal, S. L., Tilley, G. P., Sweeting, O. J.: J. Appl. Polym. Sci. *1*, 91 (1959)
112. Lindenmeyer, P. H., Lustig, S.: J. Appl. Polym. Sci. *9*, 227 (1965)
113. Keller, A., Machin, M. J.: J. Macromol. Sci. Phys. *B 1*, 41 (1967)
114. Desper, C. R.: J. Appl. Polym. Sci. *13*, 169 (1969)
115. Maddams, W. F., Preedy, J. E.: J. Appl. Polym. Sci. *22*, 2721, 2739 (1978)
116. Choi, K.-J., Spruiell, J. E., White, J. L.: J. Polym. Sci. Polym. Phys. Ed. *20*, 27 (1982)
117. Heise, B.: private communication
118. Sasaguri, K., Hoshino, S., Stein, R. S.: J. Appl. Phys. *35*, 47 (1964)
119. Oda, T., Nomura, H., Kawai, H.: J. Polym. Sci. *A* 3, 1993 (1965)
120. Kaiser, R.: Kolloid-Z. und Z. Polymere *148*, 168 (1956)
121. Kikuchi, Y., Krimm, S.: J. Macromol. Sci. Phys. *B* 4, 461 (1970)
122. Painter, P. C., Havens, J., Hart, W. W., Koenig, J. L.: J. Polym. Sci. Polym. Phys. Ed. *15*, 1237 (1977)
123. Bayer, G., Hoffmann, W., Siesler, H. W.: Polymer *21*, 235 (1980)
124. Peraldo, M.: Gazz. Chim. Ital. *89*, 798 (1959)
125. Tadokoro, H., Ukita, M., Kobayashi, M., Murahashi, S.: J. Polym. Sci. *B* 1, 405 (1963)
126. Schachtschneider, J. H., Snyder, R. G.: J. Polym. Sci. *C* 7, 99 (1964)
127. Miyazawa, T., Ideguchi, Y., Fukushima, K.: J. Chem. Phys. *38*, 2709 (1963)
128. Schmidt, P. G.: J. Polym. Sci. *A* 1, 2317 (1963)
129. Zerbi, G., Piseri, L.: J. Chem. Phys. *49*, 3840 (1968)
130. Kissin, Ju. V., Cvetkova, V. I., Cirkov, N. M.: Vysokomol. Soed. *A 10*, 1092 (1968)
131. Kotschinka, A., Grell, M.: J. Polym. Sci. *C 16*, 3731 (1968)

132. Sibilia, J. P., Wincklhofer, R. C.: J. Appl. Polym. Sci. *6*, 56 (1962)
133. Luongo, J. P.: J. Appl. Polym. Sci. *3*, 302 (1960)
134. Miyazawa, T.: J. Polym. Sci. *C 7*, 59 (1964)
135. Painter, P. C., Watzek, M., Koenig, J. L.: Polymer *18*, 1169 (1978)
136. Yamada, K., Kamezawa, M., Takayanagi, M.: J. Appl. Polym. Sci. *26*, 49 (1981)
137. Lenz, J.: Rheol. Acata *21*, 255 (1982)
138. Kakudo, M., Kasai, N.: X-Ray Diffraction by Polymers, Tokyo: Kodansha Scientific Books, 1972, p. 254
139. Zerbi, G., Gussoni, M., Ciampelli, F.: Spectrochim. Acta *23 A*, 301 (1967)
140. Samuels, R. J.: J. Macromol. Sci. Phys. *B 4*, 701 (1970)
141. Samuels, R. J.: J. Macromol. Sci. Phys. *B 8*, 41 (1973)
142. Seferis, J. C., Samuels, R. J.: Polym. Eng. & Sci. *19*, 975 (1979)
143. Samuels, R. J.: Polym. Eng. & Sci. *19*, 66 (1979)
144. Desborough, I. J., Hall, I. H.: Polymer *18*, 825 (1977)
145. Hall, I. H., Pass, M. G.: Polymer *17*, 807 (1976)
146. Hall, I. H., Pass, M. G.: Polymer *18*, 825 (1977)
147. Mencik, Z.: J. Polym. Sci. Polym. Phys. Ed. *13*, 2173 (1975)
148. Jakeways, R., Smith, T., Ward, I. M., Wilding, M. A.: J. Polym. Sci. Polym. Lett. Ed. *14*, 41 (1976)
149. Brereton, M. G., Davies, G. R., Jakeways, R., Smith, T., Ward, I. M.: Polymer *19*, 17 (1978)
150. Yokouchi, M., Sakakibara, Y., Chatani, Y., Tadokoro, H., Tanaka, T., Yoda, K.: Macromolecules *9*, 266 (1976)
151. Stambaugh, B., Koenig, J. L., Lando, J.: J. Polym. Sci. Polym. Phys. Ed. *17*, 1053 (1979)
152. Alter, U., Bonart, R.: Colloid Polym. Sci. *258*, 332 (1980)
153. Alter, U.: Ph. D. Thesis, University of Regensburg, West Germany, 1978
154. Strohmeier, W.: Ph. D. Thesis, University of Ulm, West Germany, 1981
155. Strohmeier, W., Frank, W. F. X.: Colloid & Polymer Sci. *260*, 937 (1982)
156. Ward, I. M., Wilding, M. A., Brody, H.: J. Polym. Sci. Polym. Phys. Ed. *14*, 263 (1976)
157. Stambaugh, B., Koenig, J. L., Lando, J. B.: J. Polym. Sci. Polym. Phys. Ed. *17*, 1063 (1979)
158. Tashiro, K., Nakai, Y., Kobayashi, M., Tadokoro, H.: Macromolecules *13*, 137 (1980)
159. Stambaugh, B., Koenig, J. L., Lando, J. B.: J. Polym. Sci. Polym. Lett. Ed. *15*, 299 (1977)
160. Mojsja, E. G., Menzeres, G. Ja., Miks, R.: Vysokomol. Soed. Ser. *B 24*, 441 (1982)
161. Gillette, P. C., Dirlikov, S. D., Koenig, J. L., Lando, J. B.: Polymer *23*, 1759 (1982)
162. Siesler, H. W.: J. Mol. Structure *59*, 15 (1980)
163. Siesler, H. W.: J. Polym. Sci. Polym. Lett. Ed. *17*, 453 (1979)
164. Siesler, H. W.: Makromol. Chem. *180*, 2261 (1980)
165. Holland-Moritz, K., Siesler, H. W.: Polymer Bulletin *4*, 165 (1981)
166. Stach, W., Holland-Moritz, K.: J. Mol. Structure *60*, 49 (1980)
167. Tanaka, A., Chang, E. P., Delf, B., Kimura, I., Stein, R. S.: J. Polym. Sci. Polym. Phys. Ed. *11*, 1891 (1973)
168. Holland-Moritz, K.: Kolloid-Z. Z. Polym. *251*, 906 (1973)
169. Ward, I. M.: Mechanical Properties of Solid Polymers, New York: Wiley 1971
170. Holzmüller, W.: Adv. Polym. Sci. *26*, 1 (1978)
171. Bartenev, G. M., Zelenev, Ju. V.: Physik der Polymere, Leipzig: VEB Deutscher Verlag für Grundstoffindustrie, 1979
172. Kubat, J.: Makromol. Chem. Suppl. *3*, 233 (1979)
173. Takayanagi, M.: Midland Macromol. Monograph (Meier, D. J. ed.), *4*, p. 117, London: Gordon & Breach Science Publ., 1978
174. Bonart, R., Schultze-Gebhardt, F.: Angew. Makromol. Chem. *22*, 41 (1972)
175. Flory, P. J.: Principles of Polymer Chemistry, New York: Cornell University Press, 1953
176. Andrews, E. H., Gent, A. N.: The Chemistry and Physics of Rubberlike Substances (Batemen, L. ed.), New York: Wiley, 1973
177. Treloar, L. R. G.: The Physics of Rubber Elasticity, 3rd ed., Oxford: Clarendon Press, 1975
178. Davies, C. K. L., Wolfe, S. V., Gelling, I. R., Thomas, A. G.: Polymer *24*, 107 (1983)
179. Mark, J. E., Kato, M., Ko, J. H.: J. Polym. Sci. Part C, *54*, 217 (1976)
180. Chiu, D. S., Su, T.-K., Mark, J. E.: Macromolecules *10*, 1110 (1977)
181. Eisele, U.: Progr. Colloid & Polym. Sci. *66*, 59 (1979)

182. Mark, J. E.: Polym. Eng. Sci. *19*, 254, 409 (1979)
183. Smith, T. L.: Polym. Eng. Sci. *17*, 129 (1977)
184. Flory, P. J.: Chem. Rev. *35*, 51 (1944)
185. Gee, G.: J. Polym. Sci. *2*, 451 (1947)
186. Shimomura, Y., White, J. L., Spruiell, J. E.: J. Appl. Polym. Sci. *27*, 3553 (1982)
187. Mooney, M.: J. Appl. Phys. *19*, 434 (1948)
188. Rivlin, R. S.: Phil. Trans. Royal Soc. (London), *A-241*, 379 (1948)
189. Ciferri, A., Flory, P. J.: J. Appl. Phys. *30*, 1498 (1959)
190. Mark, J. E., Flory, P. J.: J. Appl. Phys. *37*, 4635 (1966)
191. Mark, J. E.: Rubber Chem. Technol. *48*, 495 (1975)
192. Doherty, W. O. S., Lee, K. L., Treloar, L. R. G.: British Polymer J. *3*, 19 (1980)
193. Furukawa, J., Onouchi, Y., Inagaki, S., Okamoto, H.: Polymer Bulletin *6*, 381 (1981)
194. Treloar, L. R. G., Riding, G.: Proc. Roy. Soc. *A 369*, 281 (1979)
195. Siesler, H. W.: unpublished results
196. Sutherland, G. B. B. M., Jones, H. V.: Trans. Farad. Soc. *9*, 281 (1950)
197. Saunders, R. A., Smith, D. C.: J. Appl. Phys. *20*, 953 (1949)
198. Binder, J. L.: Rubber Chem. Technol. *35*, 57 (1962)
199. Binder, J. L.: J. Polym. Sci. *A 1*, 37 (1963)
200. Binder, J. L.: Appl. Spectroscopy *23*, 17 (1969)
201. Golub, M. A.: J. Polym. Sci. *36*, 523 (1959)
202. Kössler, I., Vodehnal, J.: Coll. Czechoslov. Chem. Commun. *29*, 2419 (1964)
203. Gotoh, R., Takenaka, T., Hayama, N.: Kolloid-Z. Z. Polym. *205*, 18 (1965)
204. Semon, W. L., Craig, D.: Rubber Plastic Age *40*, 140 (1959)
205. De Candia, F., Romano, G., Russo, R., Vittoria, V.: J. Polym. Sci. Polym. Phys. Ed. *20*, 1525 (1982)
206. Mitchell, J. C., Meier, D. J.: J. Polym. Sci. *A 2*, *6*, 1689 (1968)
207. Mark, J. E.: J. Polym. Sci. Macromol. Revs. *11*, 135 (1976)
208. Siesler, H. W.: unpublished results
209. Mark, J. E.: J. Chem. Ed. *58*, 898 (1981)
210. Bonart, R., Morbitzer, L., Hentze, G.: J. Macromol. Sci. *B 3*, 337 (1969)
211. Cooper, S. L., Tobolsky, A. V.: J. Appl. Polym. Sci. *10*, 1837 (1966)
212. Rinke, H.: Chimia *22*, 164 (1968)
213. Bonart, R.: J. Macromol. Sci. Phys. *B 2*, 115 (1968)
214. Clough, S. B., Schneider, N. S.: J. Macromol. Sci. Phys. *B 2*, 553 (1968)
215. Miller, G. W., Saunders, J. H.: J. Appl. Polym. Sci. *13*, 1277 (1969)
216. Morbitzer, L., Hespe, H.: J. Appl. Polym. Sci. *16*, 2697 (1972)
217. Estes, G. M., Seymour, R. W., Cooper, S. L.: Macromolecules *4*, 452 (1971)
218. Seymour, R. W., Allegrezza, A. E., Cooper, S. L.: Macromolecules *6*, 896 (1973)
219. Bonart, R., Morbitzer, L., Müller, E. H.: J. Macromol. Sci. Phys. *B 9*, 447 (1974)
220. Wilkes, G. L., Wildnauer, R.: J. Appl. Phys. *46*, 4148 (1975)
221. Samuels, S. L., Wilkes, G. L.: J. Polym. Sci. *A 2*, 11, 807 (1973)
222. Wilkes, C. E., Yusek, C. S.: J. Macromol. Sci. Phys. *B 7*, 157 (1973)
223. Wilkes, G. L., Bagrodia, S., Humphries, W., Wildnauer, R.: J. Polym. Sci. Polym. Lett. Ed. *13*, 321 (1975)
224. Wilkes, G. L., Emerson, J. A.: J. Appl. Phys. *47*, 4261 (1976)
225. Van Bogart, J. W. C., Lilaonitkul, A., Cooper, S. L.: Adv. Chem. *176*, 3 (1979)
226. Ferguson, J., Ahmad, N.: Europ. Polym. J. *13*, 859, 865 (1977)
227. Schollenberger, C. S., Dinbergs, K.: J. Polym. Sci. Polym. Symp. *64*, 351 (1978)
228. Cooper, S. L., Estes, G. M. (eds.): Multiphase Polymers, Washington, D.C.: Advances in Chemistry Series *176*, ACS 1979
229. Blackwell, J., Gardner, K. H.: Polymer *20*, 13 (1979)
230. Blackwell, J., Ross, M.: J. Polym. Sci. Polym. Lett. Ed. *17*, 447 (1979)
231. Hocker, J., Born, L.: J. Polym. Sci. Polym. Lett. Ed. *17*, 723 (1979)
232. Bonart, R.: Polymer *20*, 1389 (1980)
233. Kong, E. S. W., Wilkes, G. L.: J. Polym. Lett. Ed. *18*, 369 (1980)
234. Fridman, I. D., Thomas, E. L., Lee, L. J., Macosko, C. W.: Polymer *21*, 388, 393 (1980)
235. Lunardon, G., Sumida, Y., Vogl, O.: Angew. Makromol. Chem. *87*, 1 (1980)

236. Chan, K. W., Geil, P. H.: Polymer Communications 24, 50 (1983)
237. Blackwell, J., Nagarajan, M. R., Hoitink, T. B.: Polymer 22, 1534 (1981)
238. Born, L., Hespe, H., Crone, J., Wolf, K. H.: Colloid & Polym. Sci. 260, 819 (1982)
239. Bonart, R., Müller-Riederer, G.: Colloid & Polym. Sci. 259, 926 (1981)
240. Neumüller, W., Bonart, R.: J. Macromol. Sci. Phys. B21(2), 203 (1982)
241. Khranovskii, V. A., Gul'ko, L. P.: J. Macromol. Sci. Phys. B22 (4), 497 (1983)
242. Chamberlin, Y., Pascault, J. P.: J. Polym. Sci. Polym. Chem. Ed. 21, 415 (1983)
243. Zelenev, Ju. V., Letunovskij, M. P., Basirov, A. B.: Acta Polymerica 33, 590 (1982)
244. Kwei, T. K.: J. Appl. Polym. Sci. 27, 2891 (1982)
245. Siesler, H. W.: Polymer Bulletin 9, 382, 471 (1983)
246. Camberlin, Y., Pascault, J. P., Letoffe, M., Claudy, P.: J. Polym. Sci. Polym. Chem. Ed. 20, 1445 (1982)
247. Ferguson, J., Kumar, M.: Plastics and Rubber Proc. and Appl. 1, 259 (1981)
248. Bonart, R., Hoffmann, K.: Colloid & Polym. Sci. 260, 268 (1982)
249. Laptij, S. V., Lipatov, Yu. S., Kercha, Yu. Yu., Kosenko, L. A., Vatulev, V. N., Gaiduk, R. L.: Polymer 23, 1917 (1982)
250. Celanese Research Corp. of America, Belgian Patent 650,890 (January 23, 1965)
251. Du Pont, U.S. Patent 3,256,258 (June 14, 1966)
252. Hercules Powder Co., British Patent 1,052,550 (December 26, 1966)
253. Garber, C. A., Clark, E. S.: J. Macromol. Sci. Phys. B4(3), 499 (1970)
254. Miles, M., Petermann, Gleiter, H.: J. Macromol. Sci. Phys. B12(4), 523 (1976)
255. Knobloch, F. W., Statton, W. O.: U.S. Patent 3,299,171 (January 17, 1967)
256. Hinrichsen, G., Mießen, R., Reichhardt, M.: Angew. Makromol. Chem. 40/41, 239 (1974)
257. Quynn, R. G., Brody, H.: J. Macromol. Sci. Phys. B5(4), 721 (1971)
258. Noether, H. D.: U.S. Patent 3,513,110 (May 19, 1970)
259. Cannon, S. L., Mc Kenna, G. B., Statton, W. O.: J. Polym. Sci. Macromol. Revs. 11, 209 (1976)
260. Sprague, B. S.: J. Macromol. Sci. Phys. B8(1–2), 157 (1973)
261. Samuels, R. J.: J. Polym. Sci. Phys. Ed. 17, 535 (1979)
262. Clark, E. S.: in Structure and Properties of Polymer Films (Lenz, R. W., Stein, R. S. eds.), New York: Plenum Press, p. 267, 1973
263. Park, I. K., Noether, H. D.: Colloid & Polym. Sci. 253, 824 (1975)
264. Göritz, D., Müller, F. H.: Colloid & Polym. Sci. 253, 844 (1975)
265. Göritz, D., Müller, F. H.: Colloid & Polym. Sci. 252, 862 (1974)
266. Noether, H. D., Whitney, W.: Kolloid-Z. u. Z. Polymere 251, 991 (1973)
267. Noether, H. D.: private communication
268. Lipp, E. D., Zimba, C. G., Nafie, L. A.: Chem. Phys. Lett. 90, 1 (1982)
269. Golden, W. G., Dunn, D. S., Overend, J.: J. Catal. 71, 395 (1981)
270. Jensen, H. P.: Appl. Spectrosc. Revs. 18, 305 (1982)

H.-J. Cantow (Editor)
Received September 7, 1983

Analysis of the Fine Structure of Poly(Oxymethylene) Prepared by Radiation-Induced Polymerization in the Solid State

Yoshiaki Nakase, Isamu Kuriyama
Japan Atomic Energy Research Institute, Osaka Laboratory for Radiation Chemistry
25-1, Mii minami-machi, Neyagawa, 572 Japan

Akira Odajima
Department of Applied Physics, Faculty of Engineering, Hokkaido University
Nishi 8-chome, Kita 13-jo, Kika-ku, Sapporo, 060 Japan

In this review, the fine structure of polymers obtained by radiation-induced polymerization in the solid state, i.e., poly(tioxane) and poly(tetroxocane), is characterized by X-ray scattering, melting behavior, and radiolysis. The annealing effects on these behaviors gave quite valuable information to clarify the fine structure of the original material, since the materials used here are crystalline samples. Electron microscopic studies were also performed to confirm the suggested structure in the case of the irradiated samples.

Radiolysis reveals the regions accessible to radiation, indicating the existence of polymer chain aggregation with stress or entropy restriction accumulating somewhat periodically during the polymerization in the solid state.

Poly(tetroxocane) consists of two or three kinds of crystallites, depending on the polymerization temperature. Main- and sub-crystals are found in the polymer obtained below 80 °C, while main- and sub-crystals as well as lamellar crystals form between 80° and 90 °C, and main- and lamellar crystals above 90 °C. The sub-crystals are oriented perpendicular to the main-crystallite c-axis. Both of the sub- and lamellar crystals situate between the fibrils of the main-crystals.

Poly(trioxane) consists of two kinds of crystals, i.e., main- and sub-crystals, irrespective to the polymerization temperature.

The fibrillar crystallites of both samples, poly(trioxane) and poly(etroxocane), reveal the mean crystallite lengths of 500 Å and 300 Å, respectively.

Advances in Polymer Science 65
© Springer-Verlag Berlin Heidelberg 1984

1 Introduction

Crystalline polymers were obtained by the radiation-induced polymerization of various cyclic compounds in the solid state [1-3]. Especially the cyclic oligomers of formaldehyde $(CH_2O)_n$, i.e., trioxane (TOX) (n = 3), tetroxocane (T_EOX) (n = 4), pentoxane (P_EOX) (n = 5), and hexoxane (H_EOX) (n = 6), easily produce poly(oxymethylene (POM) by radiation-induced polymerization in the solid state [4, 5].

The relative crystal orientation of the cyclic oligomers mentioned above, i.e., monomers and their polymer products were investigated by X-ray diffraction methods. POM crystals consist of a stacking of main crystals and subcrystals. The polymer chain directions in the main crystal and the subcrystal were always oriented parallel to the specific crystallographic axis of monomer crystals.

These characteristics can be derived from a polymerization of crystalline monomers controlled topochemically. Topotactic reaction generally involve a strong correspondence between the lattices of the monomer crystal and the resulting polymer crystal. Hence, the topochemical solid state reaction occurs when sufficiently intense thermal mobility of molecules takes place in the lattices, and the distance between active centers for the polymerization in the neighboring molecules should not exceed 3.7 Å for C $\cdots$ O interactions [6]. All monomers mentioned above show quite close C $\cdots$ O interactions.

The c-axis of the main crystal of poly(oxymethylene) prepared from TOX, denoted hereafter as PTOX, coincides with that of TOX crystal, and the c-axis of the subcrystals is inclined to that of the main crystal at an angle of 76.7° [7].

On the other hand, the c-axis of POM prepared from T_EOX, denoted as PT_EOX, is parallel to the b-axis of T_EOX crystals, and the a-axis of PT_EOX is parallel to the c-axis of T_EOX.

It has been suggested [8] that PT_EOX is of the extended chain crystal, because no diffraction intensity peaks are observed by the X-ray small-angle scattering(SAXS) studies, just similar to the case of PTOX. In contrast, it has also been reported [9] that SAXS of the unannealed sample of PT_EOX revealed a long spacing of 83 Å, suggesting the existence of lamellar crystals. An electron micrograph of PT_EOX in Fig. 1.1 shows the rippled surface of the sample, indicating lamellar structure of PT_EOX crystals similar to that of the references [9, 10]. These differences may occur from the different polymerization temperature.

Morphological studies of POM prepared from TOX irradiated by α-particles showed three distinct crystals, i.e., main crystal, subcrystal, and a new twin crystal by electron microscopic and diffraction methods. These were also confirmed by annealing and melting of the crystals. This new twin crystal has not been observed in PTOX obtained by γ-ray induced polymerization of TOX.

Two molecular models of the structure of PTOX fibrils have been suggested to explain the existence of the main crystal and the subcrystal. The first model [11c] consists of both segments of extended chains and folded chains in parallel. The second model [10] consists of extended chains, but the chains are oriented differently in series, i.e., main crystal and subcrystals are inclined at an angle of 76° to the main chain.

It is important to characterize the polymers prepared in the solid state not only by X-ray diffraction but also by other methods in order to clarify their morphology

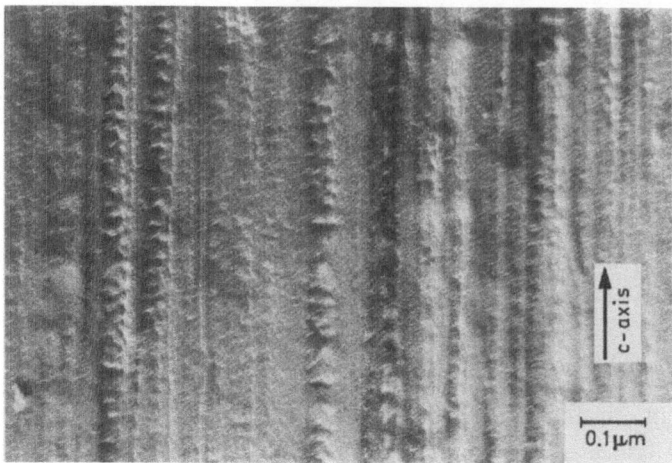

Fig. 1.1. Electron micrograph of poly(tetroxocane) prepared at 105 °C

from various stand points. For example, in the case of crystalline polymer, the investigation of the melting behavior of the polymer is one of the valuable methods. The melting point (T_m) is thermodynamically defined as $T_m = \Delta H_m / \Delta S_m$, where ΔH_m and ΔS_m denote enthalpy change and entropy change by melting, respectively. ΔH_m correlates with the molecular structure including shape and flexibility of molecules and intermolecular force, and ΔS_m with the freedom of intramolecular rotation and the orientation or regularity of molecular conformation. Therefore, the melting behavior indicates the energy of molecular aggregation in the solid state, reflecting the morphological characteristics of the crystalline polymer.

There are two thermal analytical methods for investigating the melting behavior, these being the static and dynamic methods. The dynamic method has the advantage of giving the results in a short time by increasing sample temperature, though there is a drawback of less quantitative evaluation in the results compared with the static method which keeps the sample at a constant temperature.

In this report, the dynamic methods, i.e., differential thermal analysis and differential scanning calorimetry, are applied to investigate the melting behavior of crystalline polymers, together with the conventional X-ray diffraction methods and electron microscopic study.

In order to investigate the melting behavior, the samples must be in a crystalline state, which also is convenient for studying X-ray diffraction. Poly(trioxane) and poly(tetroxocane) are selected for a prototype study. Annealing which affects the molecular arrangement within the samples, i.e., the rearrangement of the polymer chains by heating the sample below its melting points, is widely used to characterize the original structure. The annealing effects are confirmed by the melting behavior in this work [12], and structural study has been performed in detail [13]. Radiolysis of the polymeric materials is a further, suitable method to characterize the morphology [14].

2 X-ray Diffraction Behavior of Polymers

The fine structure of PTOX and PT_EOX has been investigated by small- and wide-angle X-ray scattering (SAXS and WAXS) measurements [8-10] and by thermal analysis [12a].

These structural studies have revealed that crystalline polymers such as PTOX and PT_EOX are not so perfect as the socalled single crystal, but rather disordered, on comparing with a whisker-type crystal prepared by a cationic polymerization of TOX in solution [15].

In this section, characteristics of the fine structure of their polymers are clarified from SAXS and WAXS behaviors.

2.1 SAXS and WAXS Patterns

Figure 2.1 shows SAXS patterns of PT_EOX and PTOX. PT_EOX post-polymerized at a temperature below 70 °C indicates no meridional scattering, which is similar to that of PTOX obtained at 55 °C. On the other hand, PT_EOX obtained over 80 °C shows clear meridional as well as equatorial scattering, and the spot-like scatterings also are observed outside the broad meridional scattering.

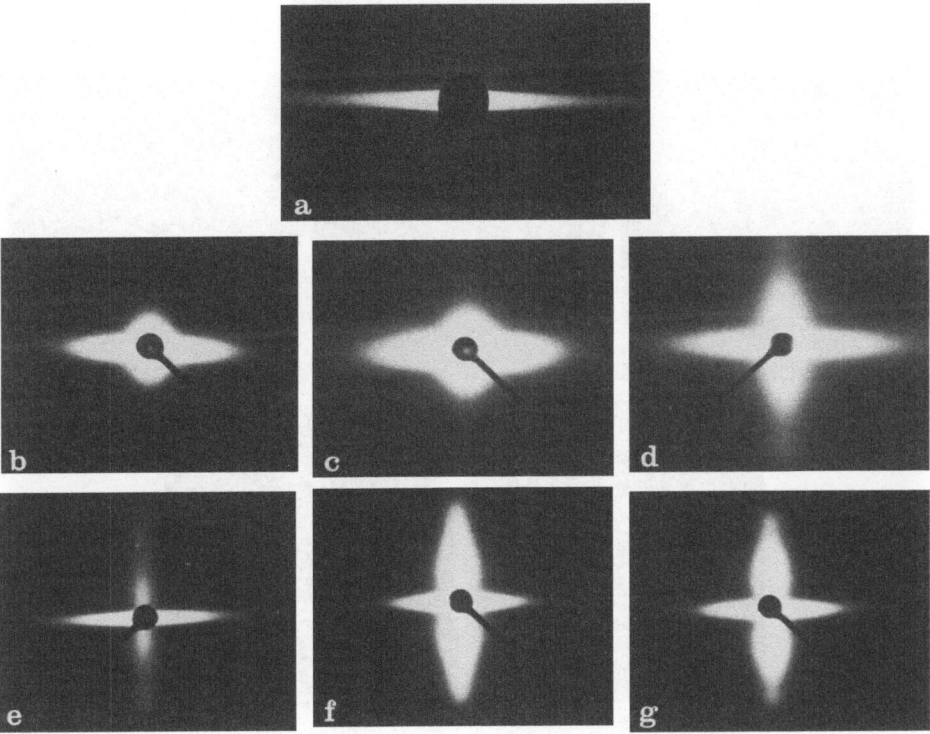

Fig. 2.1a–g. Small-angle X-ray scattering patterns of **a** poly(trioxane) prepared at 55 °C and poly-(tetroxocane) prepared at various temperatures; **b** 55 °C; **c** 70 °C; **d** 84 °C; **e** 90 °C; **f** 105 °C; **g** 110 °C. Intensities are not comparable

The secondary spot-like scattering in PT_EOX indicates the formation of lamellar stacking in fibrils of PT_EOX similar to the case described by Wegner et al. [10]. On the equator, both patterns of PTOX and PT_EOX show a strong and sharp scattering, though the equatorial streak of PTOX is sharper than that of PT_EOX.

Vertical widths on the equatorial streak of PTOX and PT_EOX were measured photographically with a microdensitometer at a desired positions of 2θ. The half widths of the streak, especially PTOX, were almost constant over the considerably wide range of 2θ. This finding indicates the possibility of the existence of needle-like voids between fibrils in the specimen with high orientation parallel to the fibre axis.

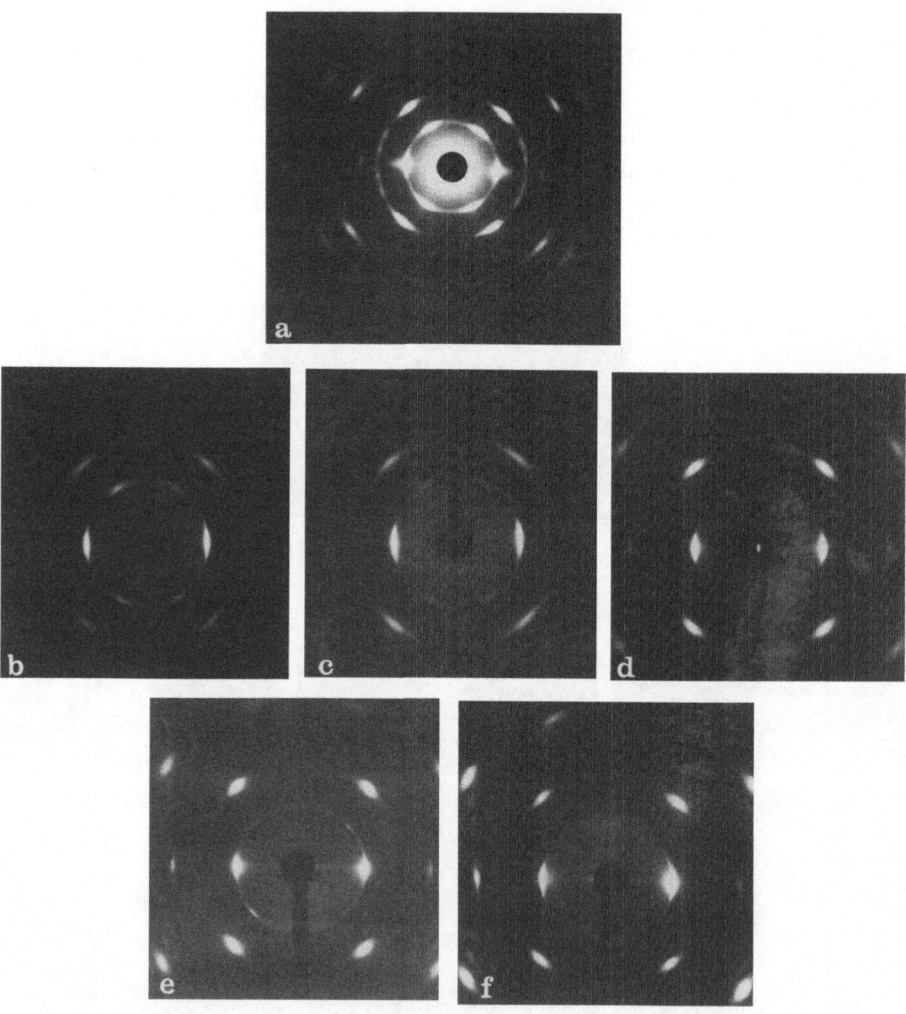

Fig. 2.2a–f. Wide-angle X-ray scattering patterns of **a** poly(trioxane) prepared at 55 °C and poly-(tetroxocane) prepared at various temperatures; **b** 55 °C; **c** 70 °C; **d** 84 °C; **e** 90 °C; **f** 105 °C. Intensities are not comparable

The longitudinal length of voids in PT_EOX was estimated to be approximately 500 Å from its half-width of meridional reflection, but the half-width of PTOX scattering could not be evaluated, the width of which attributed to the instrumental broadening. On the other hand, the void length of ca. 2000 Å for PTOX was reported from the results with a point focusing camera with a mirror and monochrometer optical system [16].

Figure 2.2 shows WAXS patterns of PT_EOX and PTOX. PT_EOX obtained at a temperature below 90 °C indicates reflections due to the sub-crystal besides those due to the main-crystals, which is similar to that of PTOX obtained at 55 °C. However, PT_EOX obtained over 90 °C shows no reflections due to the subcrystal. From the results of SAXS (Fig. 2.1) and WAXS (Fig. 2.2), it is suspected that PT_EOX obtained above 90 °C contains the main crystals and the crystals with about 100 Å spacing having the same chain direction as the main crystals. The sample prepared in the temperature range between 80° and 90 °C contains three crystals, i.e., main- and sub-crystals as well as crystals with 100 Å spacing.

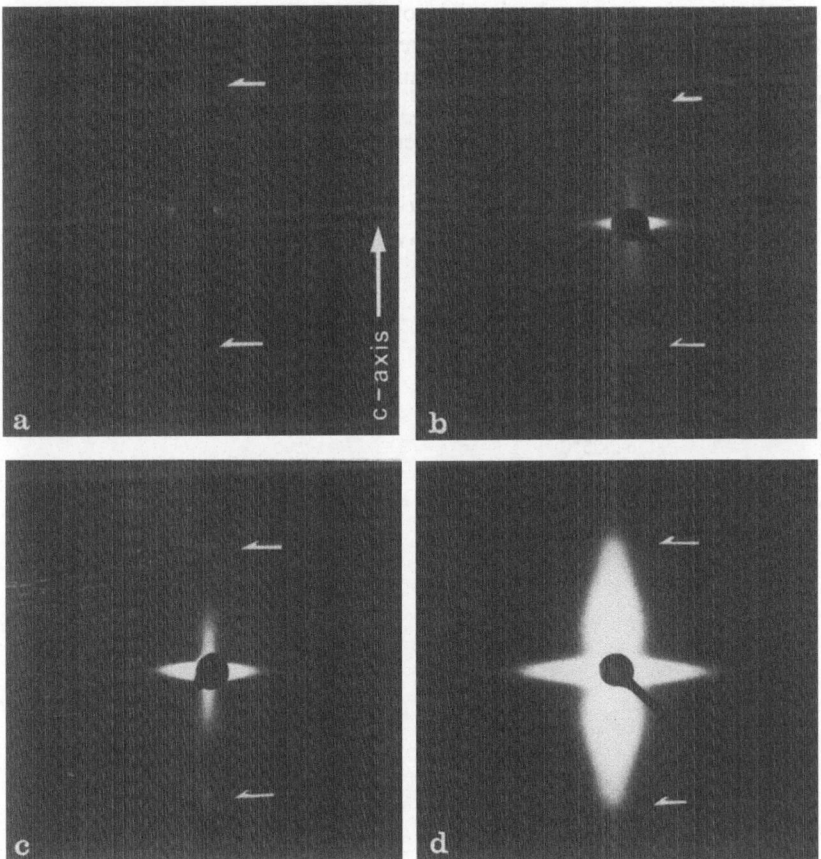

Fig. 2.3a–d. Small-angle X-ray scattering patterns of the tetroxocane and poly(tetroxocane) system: **a**, just after the polymerization; **b**, sublimed at 80 °C for 3 h under the reduced pressure; **c**, sublimed for 5 h, and **d**, sublimed for 24 h (polymer alone)

The WAXS photographs in Fig. 2.2 are obtained without rotation of the sample, but the reflection spots are found symmetrically, indicating rather random orientation of the crystals along the c-axis compared with that of PTOX, i.e., asymmetrical reflection spots [17] in the non-rotated WAXS photographs of PTOX.

2.2 SAXS and WAXS Profiles [13b]

Figure 2.3 shows the SAXS patterns for various types of PT_EOX sample systems: (a), the system includes unreacted monomeric tetroxocane just after polymerization; (b), system a is heated at 80 °C for 3 h to sublime unreacted monomer; (c), system a is heated for 5 h; and (d), system a is heated for 24 h. The sample was washed with acetone and dried to give a polymer yield of 26%. In the sample a, faint scatterings are observed in both directions of the equator and the meridian. A spot-like scattering in the meridian indicated by half-arrows is from the lamellar crystal with a Bragg spacing of ca. 100 Å, as already mentioned [13a].

The scattering intensities in both directions increase with the sublimation time (b, c, and d), indicating an increase in the electron density fluctuation by a reduction of the residual monomer. The equatorial scattering was due to the needle-shaped voids between the fibrils [18], but the mechanism for the meridional scattering corresponding to the 500 Å spacing has not been clarified yet. Therefore, the intensity curve at angles lower than 25′ in the meridional direction was investigated using a Kratky U-slit camera.

Figure 2.4 shows the SAXS curves on the meridian of the PT_EOX obtained by the post-polymerization at various temperatures with a polymer yield of about 30% in each case. A broad scattering peak was observed at about 13′ (500 Å scattering)

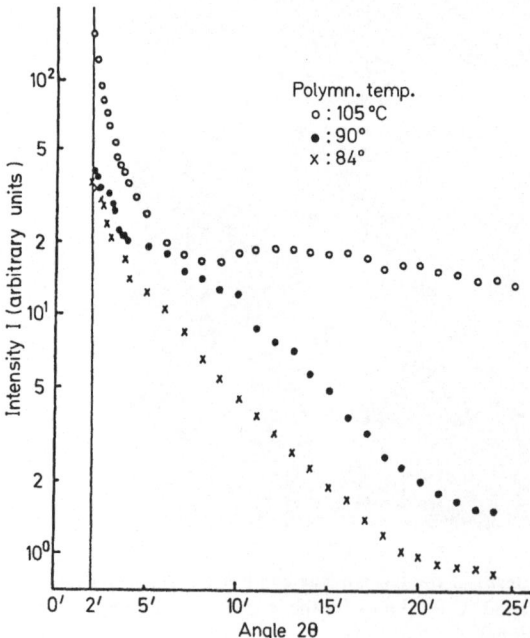

Fig. 2.4. Small-angle X-ray scattering curves (meridian) of poly(teroxocane) prepared at several different temperatures. Polymerization temperatures are indicated in the figure. Intensities are not comparable because of difference in the sample size

in the sample prepared at 105 °C, but a shoulder at about 10′ in the samples at 90° and 84 °C. The intensity curve of $PT_E OX$ prepared at 70 °C (polymer yield of about 10%) is similar to that at 84 °C, indicating a rapid decrease in the intensity with the scattering angle.

The effect of the polymer yield on the broad peak (500 Å scattering) is examined in the sample prepared at 105 °C. Figure 2.5 shows the SAXS profiles of $PT_E OX$ with various polymer yields. Since the size and weight of the sample cannot be made equal for each sample, the intensities of SAXS cannot be compared. When the polymer yield is low, the scattering maximum at about 13′ or a higher angle can hardly be observed, and as the yield increases, the maximum becomes clearer and the peak position shifts to a lower angle side.

Figure 2.6 shows the reduction of the SAXS intensity by impregnation of poly (ethylene glycol) (PEG) in the directions of both the equator and the meridian. A similar result has been obtained with a sample impregnated with silicone oil. A great decrease in the SAXS intensity by impregnation suggests that the scatterings are caused by voids formed during the polymerization crystallization.

The WAXS profiles of $PT_E OX$ obtained by post-polymerization at 81 °C and 105 °C (referred to as $PT_E OX$-81 and -105) are illustrated by the solid lines in Fig. 2.7. The 009 profiles of both samples are asymmetric with long tails on the lower-angle side. These profiles are broader than those of $PT_E OX$-62 shown by open circles in Fig. 2.7. The 0018 profiles of $PT_E OX$-81 and -105 (solid line) are somewhat symmetric but very broad, indicating that they consist of, at least, two components.

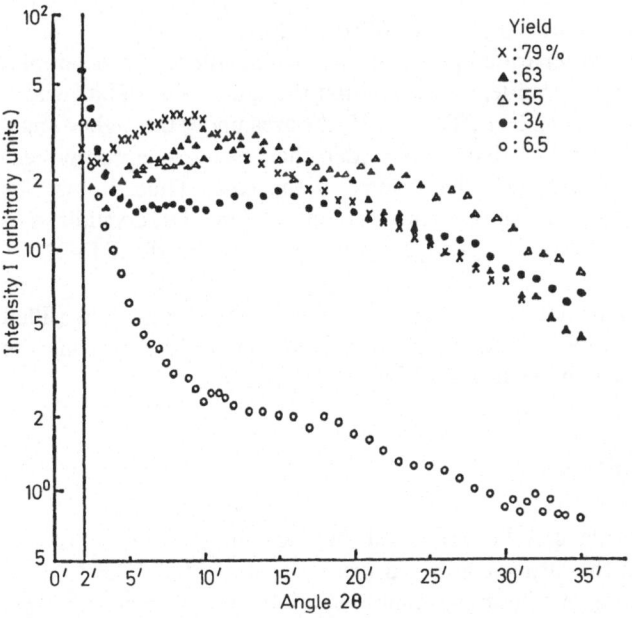

Fig. 2.5. Small-angle X-ray scattering curves (meridian) of poly(tetroxocane) polymerized at 105 °C with various polymer yields

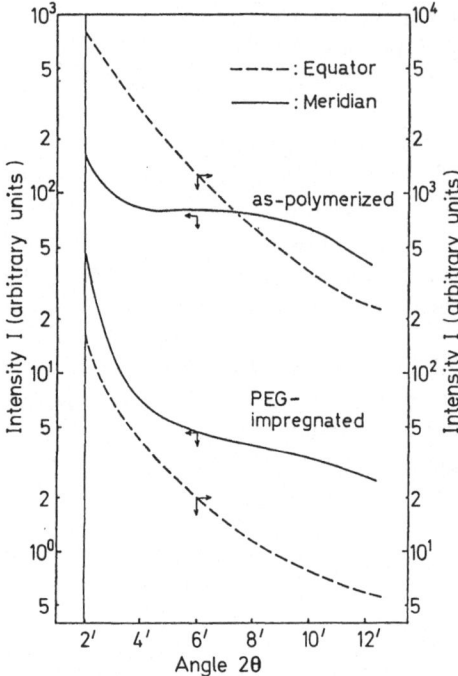

Fig. 2.6. Effect of impregnation of poly(ethylene glycol) on the small-angle X-ray scattering curves on equator and meridian of poly-(tetroxocane) polymerized at 105 °C with 40 % polymer yield

Both the 009 and 0018 profiles were decomposed into two components shown in Fig. 2.7, where the sharp profile of PT_EOX-62 (open circle) has been subtracted from the observed profiles of PT_EOX-81 and -105 (solid line), leaving a rather symmetric component with the peak at a lower-angle side (filled circles).

The distance between the two separated peaks in the 0018 profile for both samples is twice as large as that in the 009 profile, indicating that the same value of half-width can be obtained from each reflection. For $PT_EOX-105$, the area under the higher-angle component (open circles) is smaller than that under the lower-angle component (filled circles), whereas for PT_EOX-81, the reverse is the case. Thus, it may be considered that in the $PT_EOX-105$ sample the amount of lamellar crystallites is larger than that of fibrillar crystallites, and the reverse is so for the PT_EOX-81 sample.

It is very likely that the lamellar crystals give the lower-angle component (filled circles in Fig. 2.7) in the X-ray diffraction profiles, and the fibrillar crystals the higher-angle component (open circles in Fig. 2.7).

2.3 Line Broadening Analysis

A corrected diffraction profile can be expressed by the convolution [13b] of the crystallite-size profile I_s and the lattice-distortion profile I_d ($I = I_s^* I_d$). The Fourier transformation (FT) method gives the most reliable value for the 'number-average' crystallite size ($\bar{D}_n$) and the mean-squares lattice distortion ($\bar{L}_t^2$) over the distance of t.

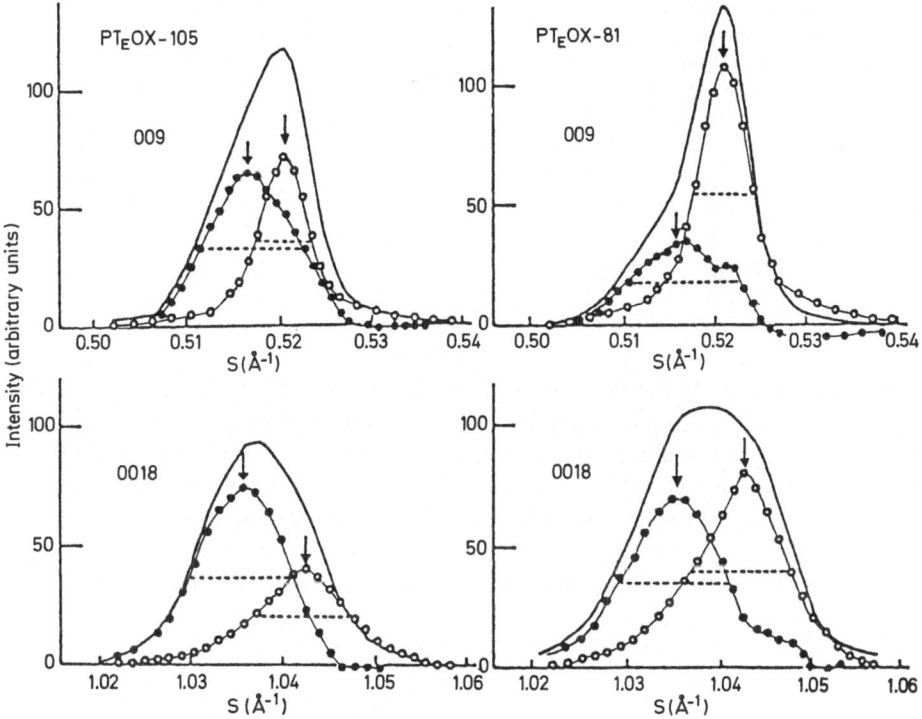

Fig. 2.7. Wide-angle X-ray scattering profiles of the 009 and 0018 reflections of poly(tetroxocane) prepared at 62 °C (open circles) and at 80 °C, 105 °C (both solid lines). The profiles of PT_EOX-81 and -105 indicate the superposition of two different profiles

According to the integral breadth (IB) method based on the paracrystalline theory [19] the IB of the 00l reflection (ΔS_l) of the Cauthy type is given by:

$$\Delta S_l = 1/D_w + (\pi g s_l)^2/\bar{c} , \tag{1}$$

where $\bar{D}_w$ denotes the 'weight-average' crystallite size along the chain axis, $\bar{c}$ the averaged c-dimension, g the longitudinal distortion parameter, and $s_l = 2 \sin \theta_l/\lambda$. Among the 00l reflections of PT_EOX only two reflections 009 and 0018 can be used, since the 0027 reflection is too weak. In the IB method, additional distortions due, for example, to lattice strains in the chain direction are neglected. In the present work, however, this method was used because of its simplicity. The experimental values of ΔS_l and s_l^2 for the 009 and 0018 reflections were substituted into Eq. (1) to evaluate g and $\bar{D}_w$.

Table 2.1 gives the sizes ($\bar{D}_w$) and the distortion parameters (g) along the chain axis for the two types of crystallites in the samples PT_EOX-62, -81, and -105. The crystal sizes of 250 Å and 100 Å are associated with the fibrillar and lamellar crystals, respectively. The distortion parameter of the fibrillar crystallites is about 3 times as large as that of the lamellar crystallite for both PT_EOX-81 and -105.

Table 2.1. Weight-average crystallite size ($\bar{D}_w$) and distortion parameter (g) in PT_EOX prepared at various temperatures

Sample	$\bar{D}_w$ (Å)	g (%)
PT_EOX-62	250	2.4
PT_EOX-81	250	2.4
	100	1.0
PT_EOX-105	250	2.4
	100	0.7

Figure 2.8 illustrates the 009 and 0018 profiles of PTOX and drawn Delrin-500. The mean crystallite size ($\bar{D}_w$) and g in both the parallel and perpendicular directions to the chain axis are estimated by using the intrinsic integral breadth of the 009 and 0018 reflections and of the four reflections of 100, 110, 210, and 300. These values are tabulated in Table 2.2 together with those of Delrin-500 for comparison. The longitudinal dimension of the main crystallite of 550 Å is extremely shorter than an extended chain length of the micro-fibril, which probably may be more than ten thousand Å [10, 14b]

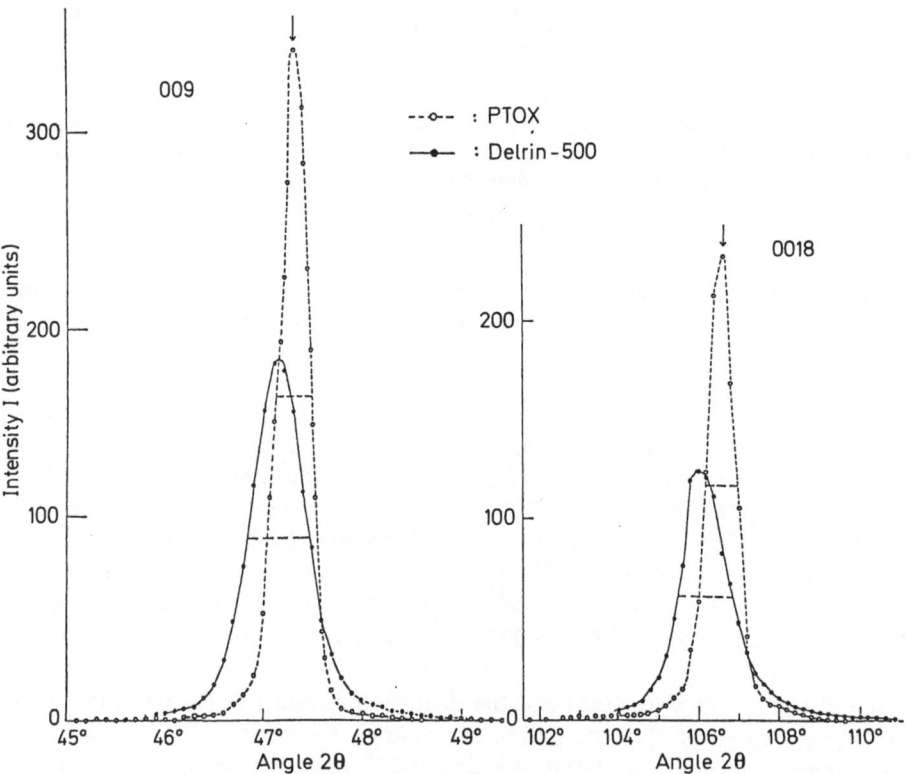

Fig. 2.8. Wide-angle X-ray scattering profiles of the 009 and 0018 reflections of poly(trioxane) as-polymerized and Delrin-500 as-drown

Table 2.2. Crystallite size ($\bar{D}_w$) and distortion parameter (g) of PTOX and Delrin-500 parallel and perpendicular to the chain axis

Sample		Parallel		Perpendicular	
		$\bar{D}_w$ (Å)	g (%)	$\bar{D}_w$ (Å)	g (%)
PTOX crystal	main-	550	0.7	300	1.4
	sub-	250	0.7	400	1.4
Delrin-500		150	1.5	150	1.8

The crystallite size obtained by the IB method defines the coherent length in the X-ray scattering. Therefore, some kinds of defects or distortions along the chain must be considered to understand such a limited coherent range. One possible explanation is given by the kinked-model of micro-fibrils [10]. The longitudinal size of the sub-crystal is 250 Å, almost half of the main-crystal size, while the transversal size of the sub-crystal seems to be a little larger than that of the main. Furthermore, the lattice distortions in both directions for both crystals are almost the same. The values of crystallite size for Delrin are close to those of the lamellar crystals. In any direction, PTOX crystals show a larger size than those of the lamellar crystallite in Delrin.

The values of the lattice constant c for various PT_EOX samples were evaluated from the 009 and 0018 reflections, and the relative deviations ($\Delta c/c$) from c = 17.29 Å determined from 009 and 17.31 Å from 0018 for PTOX are summarized in Table 2.3. The fibrillar crystals of PT_EOX give negative values of $\Delta c/c$, indicating contraction in the longitudinal direction, whereas the lamellar crystals give positive values of $\Delta c/c$, indicating expansion of the lattice. When polymerization and crystallization of PT_EOX proceed simultaneously, the molecular chains in the fibrillar crystals may incur a strong compression in a vertical direction, and the lamellar crystallites may be formed under less stressful conditions than the fibrillar crystals in PT_EOX.

Table 2.3. Relative difference in lattice constants ($\Delta c/c$) in PT_EOX-62, -81, and -105 from c = 17.29 Å determined from 009 and 17.32 Å from 0018 of poly(trioxane)

Specimen		$\Delta c/c$ (%)	
		009	0018
PT_EOX-62, Fibrils		—0.12	—0.29
PT_EOX-81, Fibrils		—0.12	—0.29
	Lamellae	0.75	0.52
PT_EOX-105, Fibrils		—0.12	—0.29
	Lamellae	0.75	0.40

3 Effects on Melting Behaviors

Since the melting behavior is related to the aggregation energy among molecular chains, morphology of polymers can be elucidated from the enthalpy study, which is a different approach from that of X-ray diffraction study.

Only a few papers, however, have been reported on the melting behavior of crystalline polymers in the relation with the polymerization condition.

3.1 Atmosphere During Polymerization

Figure 3.1 shows the DSC heating curves of PT_EOX obtained by the post-polymerization with various yields in vacuum (a) and in air (b). PT_EOX prepared in vacuum ($< 10^{-3}$ Torr) increases an endotherm at a lower temperature side with increasing polymer yield and that at a higher temperature side decreases comparatively.

The starting temperatures of the endotherm are denoted as T_{s1} and T_{s2} for the endotherms at a lower temperature side and at a higher temperature side, respective-

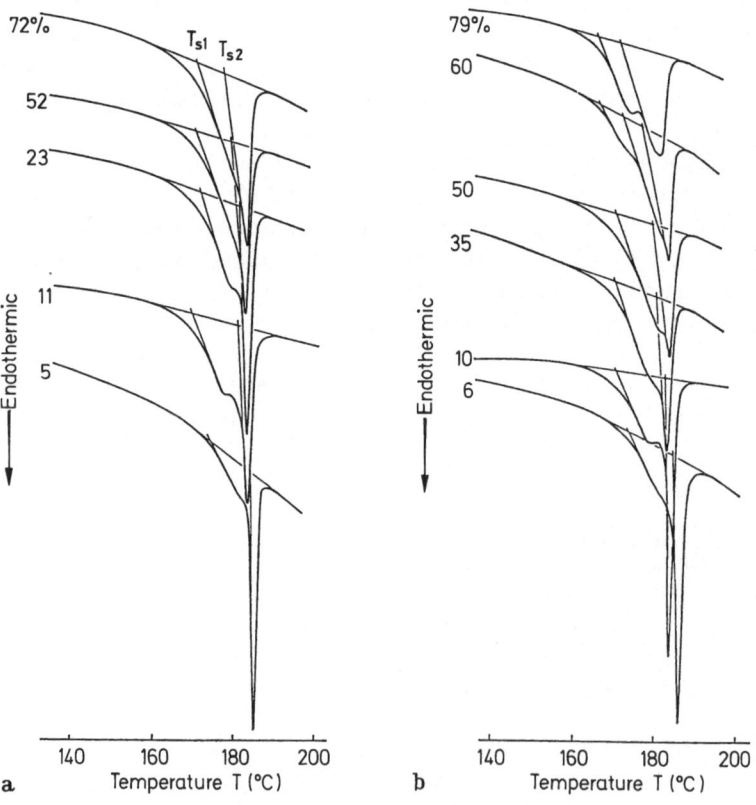

Fig. 3.1a and b. Heating curves of poly(tetroxocane) with various polymer yields: Heating·rate, 16 °C/min; polymerization at 105 °C, **a** in vacuum, and **b** in air; polymer yields are indicated in the figure

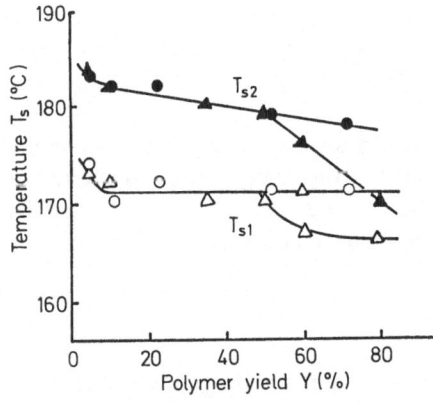

Fig. 3.2. Dependence of the starting temperatures (T_{s1} — open marks, and T_{s2} — filled marks) on the polymer yield of poly(tetroxocane): Polymerization at 105 °C, $\bigcirc$, $\bullet$, in vacuum, and $\triangle$, $\blacktriangle$, in air

ly. $PT_E OX$ prepared in air also increases the endotherm at a lower temperature, while that at a higher temperature shifts to a lower temperature side and becomes very broad. These profile changes are observed irrespective to the preirradiation dose.

Figure 3.2 shows the dependence of T_{s1} and T_{s2} on the polymer yield of $PT_E OX$. High T_{s1} and T_{s2} are observed at a polymer yields of less than 10% in both polymers obtained in vacuum and in air. T_{s1} of $PT_E OX$ prepared in vacuum (open circle) remains constant with the change of polymer yield, while T_{s2} of $PT_E OX$ prepared in vacuum (filled circles) apparently decreases with increasing polymer yield, but this tendency may be due to the superposition of two endotherms. T_{s1} and T_{s2} of $PT_E OX$ in air decrease abruptly at a polymer yield of over 50%. This fact indicates the production of polymer crystallite with a lower melting point at higher yield, reflecting the formation of low-molecular-weight polymer chains [20].

The area under the endothermic peak corresponding to the enthalpy of melting, was calculated to give almost constant values, within an experimental error irrespective to the polymer yield, that is, 205 ± 4 J/g for $PT_E OX$ obtained in vacuum, and 193 ± 4 J/g for $PT_E OX$ obtained in air. The larger value for $PT_E OX$ in vacuum than that in air corresponds to the comparatively larger endothermic area at a higher temperature side for $PT_E OX$ in vacuum, since the enthalpy of melting of a crystal with a lower melting temperature is smaller than that with a high melting point [21].

The effect of the atmosphere on the morphology of PTOX is mainly attributed to the polymer chain length, that is, polymer crystals with a high melting point can be formed at the beginning of the polymerization, but their melting point decreases with increasing polymer yield in $PT_E OX$ prepared in air, indicating chain scission occuring during polymerization [20].

3.2 Preirradiation Dose

$PT_E OX$ prepared at various preirradiation doses in the range between 0.1 and 30 kGy shows double endothermic profiles, though T_{s1} and T_{s2} decrease rather significantly with increasing dose, indicating a comparatively larger area of the endotherm at a lower temperature side.

By increasing the preirradiation dose, the amount of active centers increases to accelerate the termination as well as the propagation reaction and to form low-molecular-weight polymer chains [22], which produce a polymer crystal with a lower melting point.

The effects of atmosphere during polymerization and of preirradiation dose are rather clear in the melting behavior, but not clear in the X-ray diffraction behavior mentioned in Section 2. The orientation of the polymer chains is not affected by atmosphere nor preirradiation dose; only the polymer chain length is affected.

3.3 Polymerization Temperature

Figure 3.3 shows the heating curves of PT_EOX obtained in air; the polymer yields are limited to ca. 10%. PT_EOX obtained at 55°, 70°, and 84 °C shows a single endothermic peak with broad onset at a lower temperature side of approximately 170 °C. PT_EOX obtained at a temperature above 90 °C shows a double endotherm with a different profile from that of PT_EOX obtained at a temperature below 90 °C. On increasing the polymerization temperature above 90 °C, the starting temperature of the melting (T_{s1}) shifts to a lower temperature side as shown in Table 3.1, indicating the formation of polymer crystallite with a low melting point.

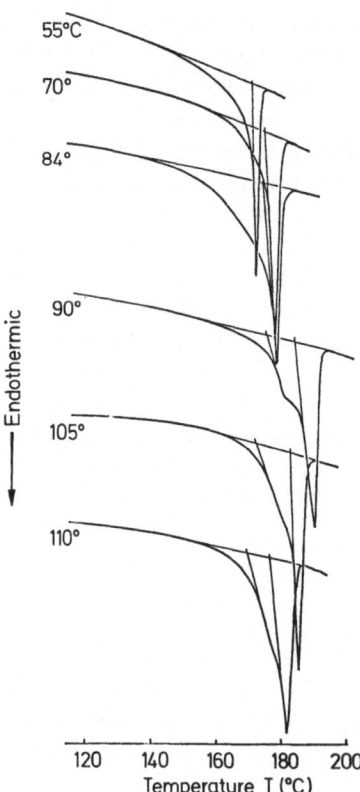

Fig. 3.3. Heating curves of poly(tetroxocane) prepared at various temperatures: Polymer yields, $10 \pm 3\%$; hesting rate, 16 °C/min; polymerization temperatures are indicated in the figure

Table 3.1. Area under the endothermic peak in the heating curves of poly(tetrox-ocane) prepared at various temperatures (from Fig. 3.3)

Polymerization temperature (°C)	Melting point (°C)		Sample weight (mg)	Area (J/g)
	T_{s1}	T_{s2}		
55	—	170	1.08	195
70	—	177	1.00	200
84	—	176	1.08	210
90	176	185	1.01	215
105	174	183	1.13	205
110	173	179	1.30	205

Table 3.1. also shows some area under the endothermic peak (Fig. 3.3) of PT_EOX obtained at various temperatures. The area, or the enthalpies of melting, remain constant in all samples, indicating no special change in the polymer chain aggregation at the first stage of the polymerization at various temperatures (polymer yield ca. 10%).

At the later stage of polymerization, the enthalpy of melting decreases slightly by forming crystallites with a lower melting point, as mentioned above.

3.4 Molecular Weight at a Constant Polymer Yield [12a]

Figure 3.4 shows that when the molecular weight of polymer increases, the onset and the peak of the endotherm shift to higher temperatures and the profile of the double melting endotherm becomes clearer. It has already been determined that the molecular weight distribution becomes broad showing the increase of the higher molecular weight fraction when the average molecular weight of the sample increases [23]. The profile of the melting endotherm, therefore, is explicable in terms of the molecular weight and its distribution of polymer used.

When the polymer yield is less than about 15%, the profile of the endotherms of the polymer obtained by the polymerization in air and in vacuum is almost the same, provided that the polymer has the same polymer yield and molecular weight.

Figure 3.4 is the case of the low polymer yield of 11%, and comparison has been made with polymers obtained in air and in vacuum at the same time. In the case of the polymer yields of 25% or 40%, the same result as in Fig. 3.4 was obtained with polymers formed in vacuum.

3.5 Polymer Yield at a Constant Molecular Weight [12a]

For the case of a molecular weight of 2×10^5 Fig. 3.5 shows the melting endotherms of polymer obtained in the absence of oxygen with various polymer yields of 10%, 27%, and 37%, respectively.

A double melting endotherm is observed, the endotherms apparently being converted into a single but broad peak when the polymer yield increases. However, the onset of the endotherm and the peak temperature of the total endotherm move

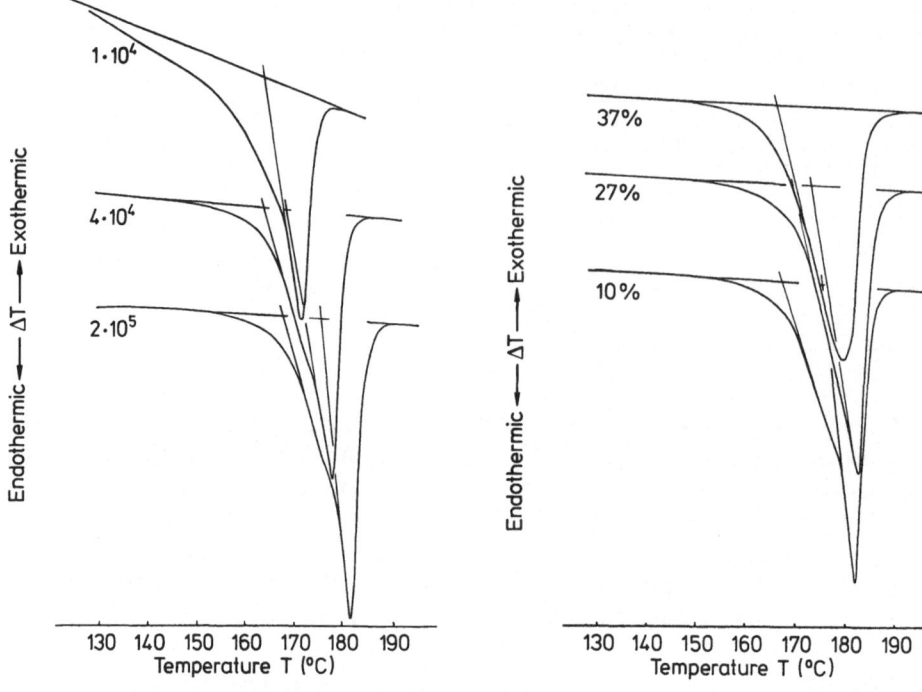

Fig. 3.4. Heating curves of poly(tetroxocane) obtained with various molecular weights at the same polymer yield (11 %): heating rate, 20 °C/ min; sample weight, 0.37 ± 0.01 mg; molecular weights are indicated in the figure

Fig. 3.5. Heating curves of poly(tetroxocane) obtained at various polymer yields with the same molecular weight (I) (2×10^5): heating rate, 20 °C/min, polymer yields are indicated in the figure

only slightly to a lower temperature. The same result is obtained as shown in Fig. 3.6 when the molecular weight is extremely small (1×10^4). This phenomenon may be understood by the relative decrease of the high temperature peak.

As the polymer yield is low, the onset of the melting endotherm is at low temperature, whereas with the increasing polymer yield, the onset moves to a higher temperature. Such a change of profile suggests that the polymer crystal formed becomes uniform, or it contains tightly packed polymer chains, when the polymer yield increases.

The experimental observation is summarized as follows.

The melting endotherm of PT_EOX shows a double peak. When the molecular weight of the polymer decreases, the onset and the peak temperatures of the endotherm shift to lower temperatures, while the profile of the endothermic peak only changes slightly. When the polymer yield increases, the lower temperature peak of a double endotherm becomes intense to give a broad profile.

On considering these observations, the difference in the melting endotherms between C and D in Fig. 3.6 is noticed at the onset of the endotherm. From the viewpoint of polymer yield, the onset of the endotherm D should be at a temperature higher than that of C. Judging from the endotherm A, the onset of D is that of a

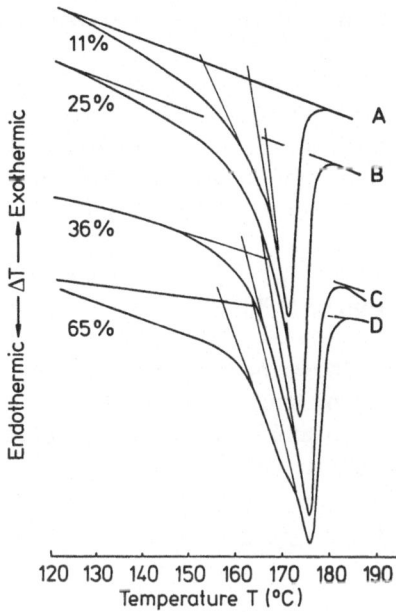

Fig. 3.6. Heating curves of poly(tetroxocane) obtained at various polymer yields with the same molecular weight (II) (1×10^4): heating rate, 20 °C/min, polymer yields are indicated in the figure

polymer with a low polymer yield. It has already been found that the polymer giving the melting endotherm D has a rather narrow molecular weight distribution [23]. Therefore it can be assumed that the polymer crystal formed during the polymerization in air followed by the degradation has the same nature as that formed at the early stage of the polymerization, and that it cannot be of the tight packing of the polymer chains. It is concluded that the profile of the melting endotherm is dependent on the molecular weight and the molecular weight distribution, which affect the morphology of the polymer formed, i.e., the aggregation of the polymer chain, crystallite size and perfection. It is evident that the morphology of the polymer has been changed during the polymerization in air.

In other words, the degradation of the polymer chain during the polymerization in the presence of oxygen occurs in the matrix of the polymer formed.

In conclusion of the melting behavior of PT_EOX, the endothermic profiles of PT_EOX prepared below 90 °C is different from that prepared above 90 °C. From the results of the X-ray diffraction, mentioned in Section 2, can be seen that the polymerization of T_EOX in the solid state produces the main and sub-crystals below 90 °C, while the main crystal and a crystal corresponding to 100 Å scattering in SAXS are formed above 90 °C. Chatani et al. [8] used PT_EOX prepared at 62 °C to show main and sub-crystals from WAXS study, and Amano et al. [9] used PT_EOX prepared at 100 °C to show only lamellar type crystals from SAXS and EM studies. Both results seem to be somewhat similar, but we can not compare their data since the characteristics of PT_EOX prepared at different temperatures are different. PT_EOX prepared at 100 °C contains the main crystals beside lamellar type crystals as mentioned later.

The morphology of the polymers is investigated in detail by the annealing method and by radiolysis with γ-irradiation.

4 Annealing Effects on the Melting and X-ray Scattering Behaviors

It is well-known that POM has a very low thermal stability, probably as it is readily oxidized upon heating. Thus, the atmosphere during annealing affects the micro (fine) structure of the polymer crystal. For example, the subcrystals disappear when PTOX is heated to 170 °C for 2 h in air. Therefore, the annealing condition must be specified in order to investigate the precise morphological change of the sample.

4.1 Annealing in Nitrogen or Oxygen Atmosphere [12b)]

PT_EOX is used here as a typical example. Figure 4.1 shows the melting endotherm in the heating curve of PT_EOX heated from the annealing temperature of 170 °C (C), as well as heating curves of an original sample (A) and a similar sample once cooled down to room temperature after annealing (B) for comparison.

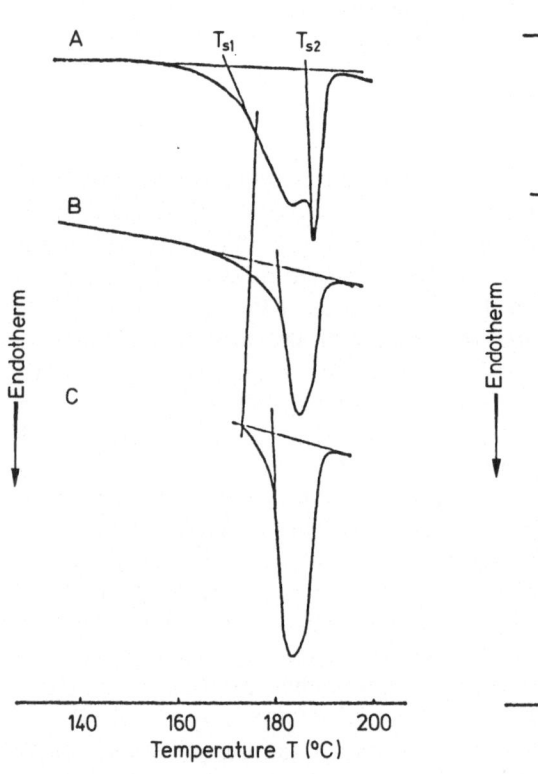

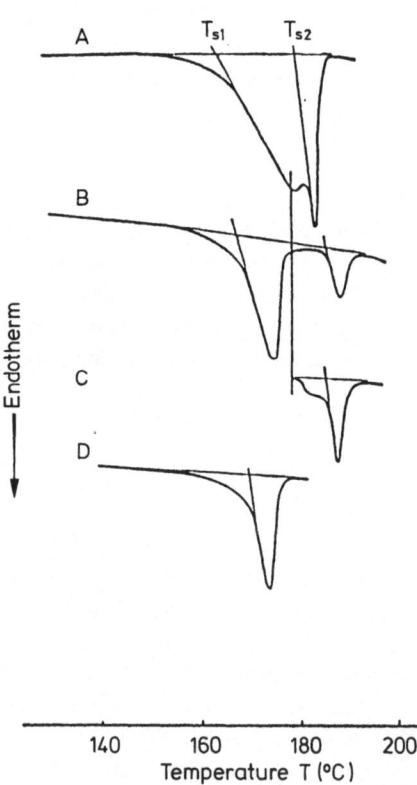

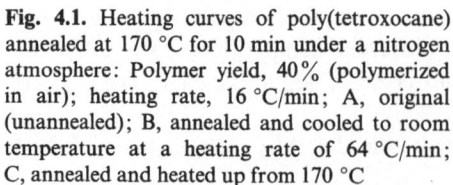

Fig. 4.1. Heating curves of poly(tetroxocane) annealed at 170 °C for 10 min under a nitrogen atmosphere: Polymer yield, 40% (polymerized in air); heating rate, 16 °C/min; A, original (unannealed); B, annealed and cooled to room temperature at a heating rate of 64 °C/min; C, annealed and heated up from 170 °C

Fig. 4.2. Heating curves of poly(tetroxocane) annealed at 178 °C for 10 min under a nitrogen atmosphere: Same sample as in Fig. 4.1; heating rate, 16 °C/min; A, original; B, annealed and cooled to room temperature at a rate of 64 °C/min; C, annealed and heated up from 178 °C; D, melt-crystallized from 195 °C at a rate of 64 °C/min

The area under the endothermic peak of the curve C coincides fairly well (in: mJ · mg^{-1}) with that of curve B. During annealing at 170 °C rearrangement of the polymer chains occurs, but no partial melting is detectable.

Figure 4.2 shows the heating curves of the sample annealed at 178 °C, The heating curve of the sample annealed for 10 min and heated up from 178 °C (C) without intermediate cooling has a quite sharp, but small, endothermic peak. The area under the peak of curve C nearly equals the area under the endothermic peak at the higher temperature of curve B of the sample cooled once to room temperature.

The melting endotherm at a lower temperature of curve B is quite similar to that of curve D of the sample cooled down to room temperature after complete melting. It has been confirmed that the endotherm at a lower temperature in the heating curve of the sample annealed at 178 °C is of the crystal formed from the partially molten phase.

The effect of oxygen during annealing has been investigated.

Figure 4.3 shows the melting endotherms of the sample heated up from the annealing temperature (B and D) without intermediate cooling as well as the melting endotherms of the sample cooled down once to room temperature (A and C). The endothermic peak observed in the heating curves of samples A and B is due to the melting of the polymer crystal presented in the system during annealing at 170 °C. The heating curves of samples C and D show that little of the crystal is left during annealing at 176 °C for 5 min under oxygen stream and that the melting endotherm of curve C is quite similar to that of the melt-crystallized sample (E), i.e., the second run of heating curve C.

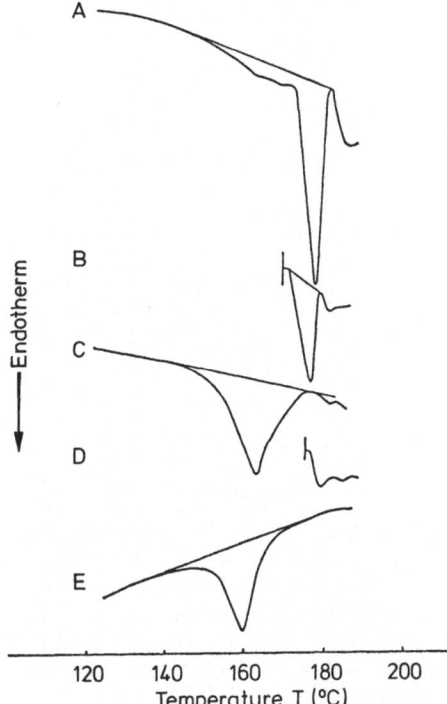

Fig. 4.3. Heating curves of poly(tetroxocane) annealed under an oxygen atmosphere: Same polymer as in Fig. 4.1; heating rate, 16 °C/min; A, annealed at 170 °C for 10 min and cooled to room temperature at a rate of 64 °C/min; B, annealed and heated up from 170 °C; C, annealed at 176 °C for 5 min and cooled to room temperature at 64 °C/min; D, annealed and heated up from 176 °C; E, melt-crystallized (2nd run of C)

The comparison between Fig. 4.2 and 4.3 suggests: In the annealing at 170 °C rearrangement of the polymer chains occurs both in a nitrogen and an oxygen atmosphere. In the annealing above 170 °C, namely at 176 °C or 178 °C, partial melting occurs merely leaving the main crystals in the case of a nitrogen atmosphere. In the case of an oxygen atmosphere, main crystals are left during annealing, and the melting endotherm at the lower temperature is quite broad at a low temperature, indicating chain scission during annealing.

It has been confirmed that the degradation of POM occurs easily when the sample is in the molten phase in the presence of oxygen.

4.2 Annealing under Nitrogen Atmosphere in Relation to the Starting Temperature of the Melting [12b)]

Upon annealing of PT_EOX at 170 °C, the rearrangement of the polymer chains occurs to give a single melting endotherm (Fig. 4.1), while at 178 °C the double melting endotherm is observed (Fig. 4.2). Both temperatures, 170 °C and 178 °C, are higher

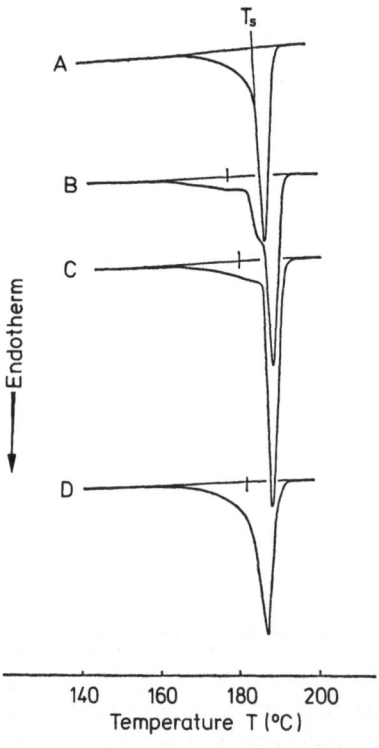

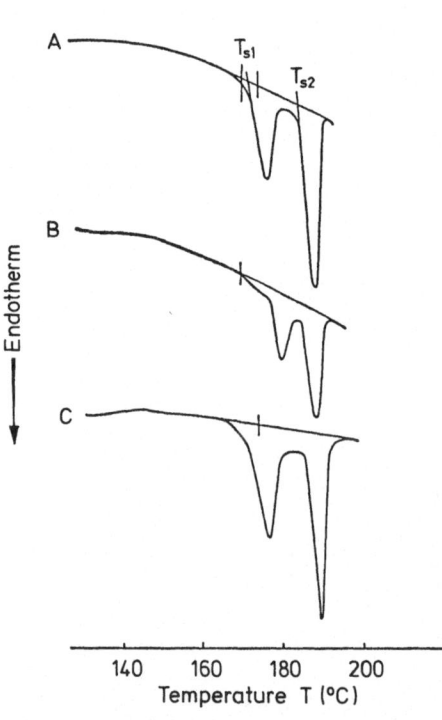

Fig. 4.4. Heating curves of the melt-crystallized poly(oxymethylene), acetylated poly(trioxane), annealed for 10 min at various temperatures under a nitrogen atmosphere: Heating rate, 16 °C/min; A, original; B, annealed at 165 °C (T_s − 5.5); C, annealed at 168 °C (T_s − 2.5); D, annealed at 171 °C (T_s + 0.5)

Fig. 4.5. Effect of the annealing temperature (in nitrogen) on the heating curves in the case of annealing below the starting temperature of the endotherm (T_{s1}): Heating rate, 16 °C/min; A, annealed at 178 °C for 5 min; B, sample A annealed again at 170 °C for 30 min; C, sample A annealed again at 174 °C for 10 min

than the starting temperature of an endotherm at a lower temperature (T_s) of the original sample as shown in Fig. 4.1 and 4.2. Therefore, a preferential temperature must be defined for the annealing effect on PT_EOX, as described elsewhere [12a]. The effect of an annealing temperature on the endothermic profiles is examined in relation to the starting temperatures of melting (T_{s1} and T_{s2}) with melt-crystallized POM, PT_EOX, and melt-crystallized PT_EOX.

Figure 4.4 shows the heating curves of the melt-crystallized POM [acetylated poly(trioxane) by acetic anhydride]. In the annealing at a temperature below T_s (170.5 °C) the rearrangement of the original polymer chains occurs to give a melting endotherm (B and C) at a temperature higher than that (A) of the original, while the annealing above T_s makes no spacial change in the melting endotherm (D) compared with that of the original.

Figure 4.5 shows the heating curves of PT_EOX annealed at 178 °C for 5 min (A), and subsequently annealed at 170 °C for 30 min (B) or at 174 °C for 10 min (C). In the annealing at a temperature (170 °C) below T_{s1} (172 °C) of the starting sample the endotherm at the lower temperature of the heating curve A shifts to a higher temperature as shown in curve B, while the endotherm at the higher temperature is kept constant. On the other hand, in the annealing above T_{s1} the profiles of the heating curve are quite similar to each other (A and C). These facts indicate that the structure corresponding to the endotherm at the lower temperature has a striking resemblance to that of the melt-crystallized material shown in Fig. 4.4.

Figure 4.6 shows the heating curves of the sample annealed at 174 °C for 10 min (A), subsequently annealed again at 178 °C for 15 min (B), and annealed thoroughly at 178 °C for 15 min (C).

In the annealing of the starting sample above T_{s1} and below T_{s2}, or 178 °C, the endotherm at the lower temperature remains almost constant, whereas that at the higher temperature shifts even higher (B). The area under the total endotherm remains almost constant. However, the profile of curve B is quite different from that of curve C of the sample prepared from as-polymerized material annealed at 178 °C for 15 min. These facts indicate that the annealing of PT_EOX at 174 °C causes the reorganization of the crystallites and the resulting structure is harder to anneal than that of the as-polymerized.

Figure 4.7 shows the heating curves of the melt-crystallized PT_EOX annealed at various temperatures. In order to prepare the melt-crystallized form, the as-polymerized PT_EOX is kept at about 195 °C for 1 min after melting, and then cooled at 64 °C/min. The heating curve of this sample (A) gives a single endotherm. In the annealing at each temperature, a double endotherm is observed in the heating curves.

The endotherm at the lower temperature increases and shifts to the higher temperature with increasing annealing temperature, while an endotherm at the higher temperature decreases but shifts at still higher temperatures. A single endotherm has been observed in the heating curve of the sample heated up directly from 170 °C ($T_{s1} + 1$ °C) without cooling, which indicates the presence of the polymer crystal in the system during the annealing of the melt-crystallized PT_EOX.

These facts cannot be supposed in the melt-crystallized POM in Fig. 4.4, and in the structure corresponding to the endotherm at lower temperature as shown in Fig. 4.5.

A similar result was reported in the melt-crystallized polyethylene [24], where the

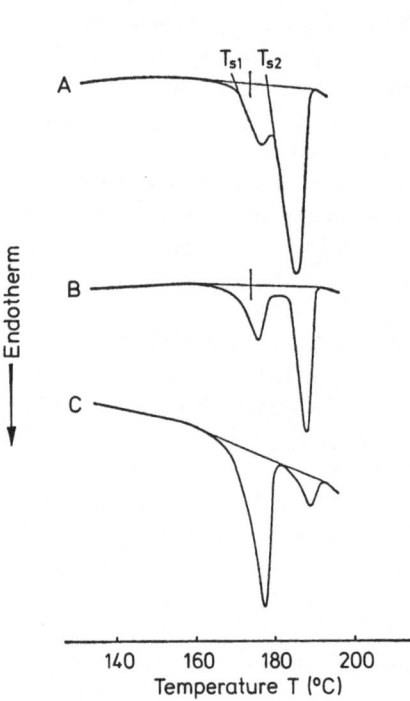

Fig. 4.6. Effect of the annealing temperature (in nitrogen) on the heating curves in relation to T_s of poly(tetroxocane): Heating rate, 16 °C/min; A, annealed at 174 °C for 10 min; B, sample A annealed again at 178 °C for 15 min; C, annealed at 178 °C for 15 min

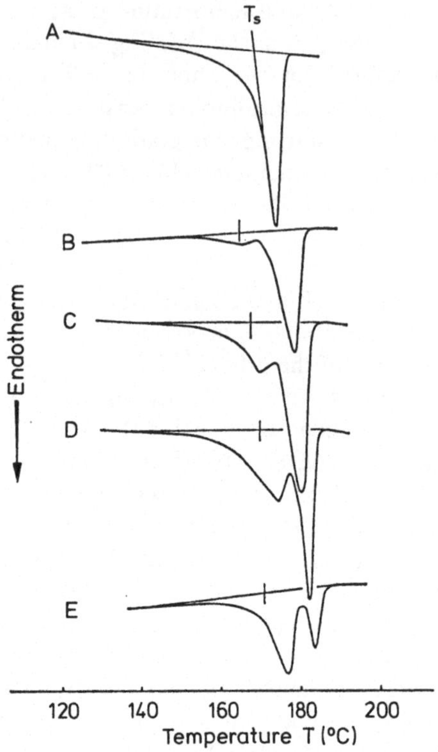

Fig. 4.7. Effect of the annealing temperature (in nitrogen) on the heating curves in relation to T_s of the melt-crystallized poly(tetroxocane) at a cooling rate of 64 °C/min: A, original melt-crystallized; B, sample A annealed at 165 °C ($T_s - 4.0$); C, sample A annealed at 168 °C ($T_s - 1.0$); D, sample A annealed at 170 °C ($T_s + 1.0$); E, sample A annealed at 171 °C ($T_s + 2.0$)

molecular weight fractionation process occurred during annealing. Accordingly, such a fractionation is assumed in the case of PT_EOX in Fig. 4.7.

These phenomena may reflect the structure of the as-polymerized PT_EOX. Namely two kinds of aggregations are expected.

4.3 Small- and Wide-Angle X-ray Scatterings

Changes in small-angle X-ray scattering (SAXS) patterns were caused by annealing [12b]. The feature of SAXS patterns of the annealed sample shows the discrete scatterings in the meridional direction indicating the existence of electron density fluctuation (with a long spacing) along the longitudinal direction, while all the samples reveal a separated endotherm in the heating curve at the lower temperature in addition to that at the higher one. By increasing the annealing time or temperature, the discrete scatterings become clearer and finally spot-like corresponding to the long spacing (ca. 170 Å) of the melt-crystallized material. This fact suggests the realization of the partial melting during annealing and the recrystallization upon cooling. The diffuse equatorial scattering decreases with the increase of the melting endotherm at the lower temperature by annealing.

Since it was reported [18] that the diffuse equatorial scattering in SAXS is caused by the microvoids between the fibrillar crystals, the lamellar crystal formed from the partial melt may fill up the microvoids so that the diffuse scattering in the equatorial direction decreases.

The lamellar crystal has a preferential orientation along the fibrillar direction (the c-axis of polymer crystal). Thus, it is assumed that the overgrowth of the folded chain crystal (shish kebab type) takes place in the formation of the lamellar crystal formed from the partial melt upon cooling.

The SAXS patterns of the melt-crystallized PT_EOX shows only a ring-like diffuse scattering to indicate no preferential orientation of the lamellar crystal formed from the molten phase.

Therefore, it is suspected that the fine structure of the as-polymerized PT_EOX is different from the melt-crystallized (the lamellar crystal), since the rearrangement of the polymer chains of the as-polymerized by annealing takes place even above the starting temperature (T_s) of the melting.

The SAXS intensity curve of the as-polymerized PT_EOX obtained by a Kratky U-slit camera shows several intensity maxima. Two maxima are rather clear corresponding to the long spacings of ca. 100 Å and ca. 500 Å, respectively.

4.4 100 Å and 500 Å Spacings of Poly(tetroxocane) [13a]

In order to clarify the characteristics of the 100 Å spacing in SAXS patterns of the sample prepared at a temperature above 90 °C, annealing effect on the SAXS and the melting behaviors of PTOX are examined.

Figure 4.8 shows the SAXS patterns of the as-polymerized (a), the annealed (b–g), and the melt crystallized samples (h). In the annealing at 140 °C (b) the 100 Å scattering remains unaltered, while the intensity maximum of the 500 Å scattering shifts to the center. In the annealing above 150 °C (c–f) both maxima shift to the center. The SAXS pattern of the sample annealed at 181 °C (g) is quite different from the others. That is, the equatorial scattering is very faint, indicating the diminution of the microvoids, and the weak scattering which is equivalent to the Bragg spacing (190 Å) of the melt-crystallized form [a faint ring-like scattering in the sample (h)] is also observed together with the intense scattering corresponding to 280 Å. Figure 4.9 shows WAXS patterns of the as-polymerized (a), the annealed (b and c), and the melt-crystallized PT_EOX (d). The as-polymerized form (a) shows a highly preferential orientation of the polymer chains along the c-axis and no twin reflection, since the polymerization temperature is 105 °C. The patterns of the annealed samples (b and c) are similar to that of the as-polymerized, though the sample annealed at 181 °C shows the double melting endotherm in the heating curve, as described later. On the other hand, the melt-crystallized form (d) shows no preferential orientation of the polymer chains. An amorphous halo inside the crystal

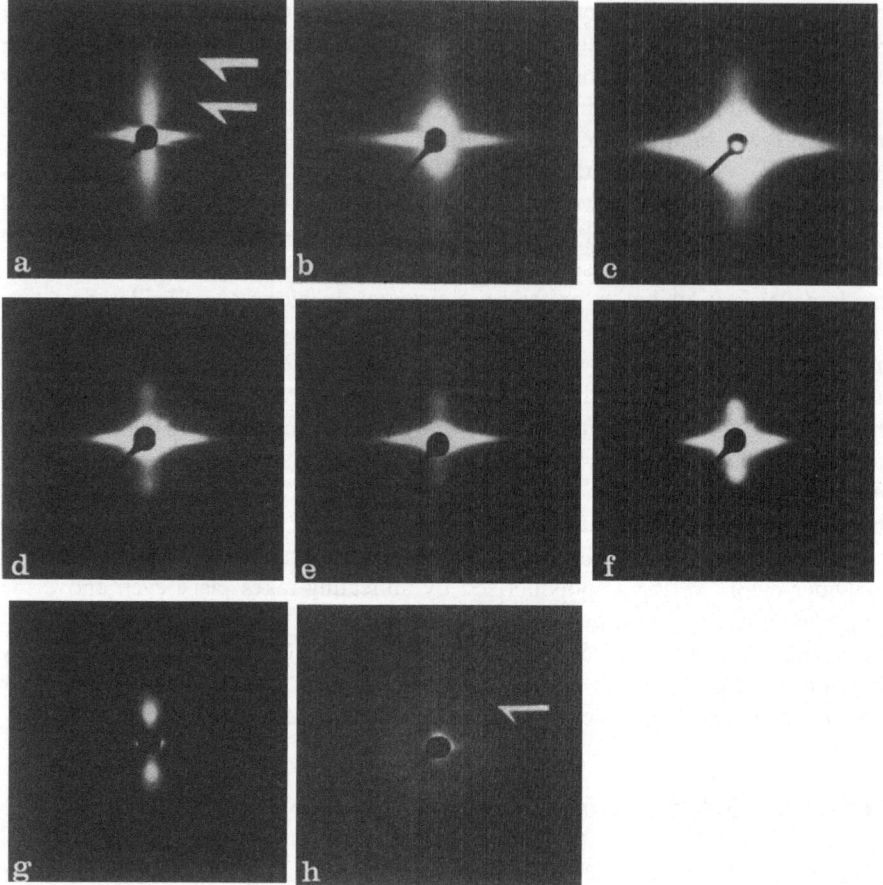

Fig. 4.8a–h. Small-angle X-ray scattering patterns of poly(tetroxocane) annealed in nitrogen: **a** original; and annealed, **b** at 140 °C for 30 min; **c** 150 °C, 30 min; **d** 168 °C, 10 min; **e** sample **d** annealed at 171.5 °C for 15 min; **f** sample **e** annealed at 176 °C for 12 min; **g** 181 °C, 2 min; **h** 195 °C, 1 min (melt-crystallized). Intensities are not comparable

100 reflection peak is detected if one closely inspects the WAXS photographs of all samples.

Figure 4.10 shows the heating curves of the as-polymerized (A), the annealed (B ∼ G), and the melt-crystallized samples (H). The samples (B ∼ D and G) are annealed directly at the desired temperature, and E and F are annealed stepwise, since the annealing at temperatures below T_s causes the rearrangement of the polymer chains as mentioned in Section 4.2.

As shown in Fig. 4.10, the heating curves of the as-polymerized sample (A) shows the double melting endotherms, so the starting temperature of the melting endotherm at a lower temperature yields T_{s1}, and that at a higher temperature T_{s2}. The samples (B ∼ F) are annealed at a temperature below T_{s1} of each starting material. The annealing at temperatures above T_{s1} causes partial melting leading to a double melting endotherm (G) in the heating curve, in which the endotherm at the lower temperature is similar to that of the melt-crystallized form (H).

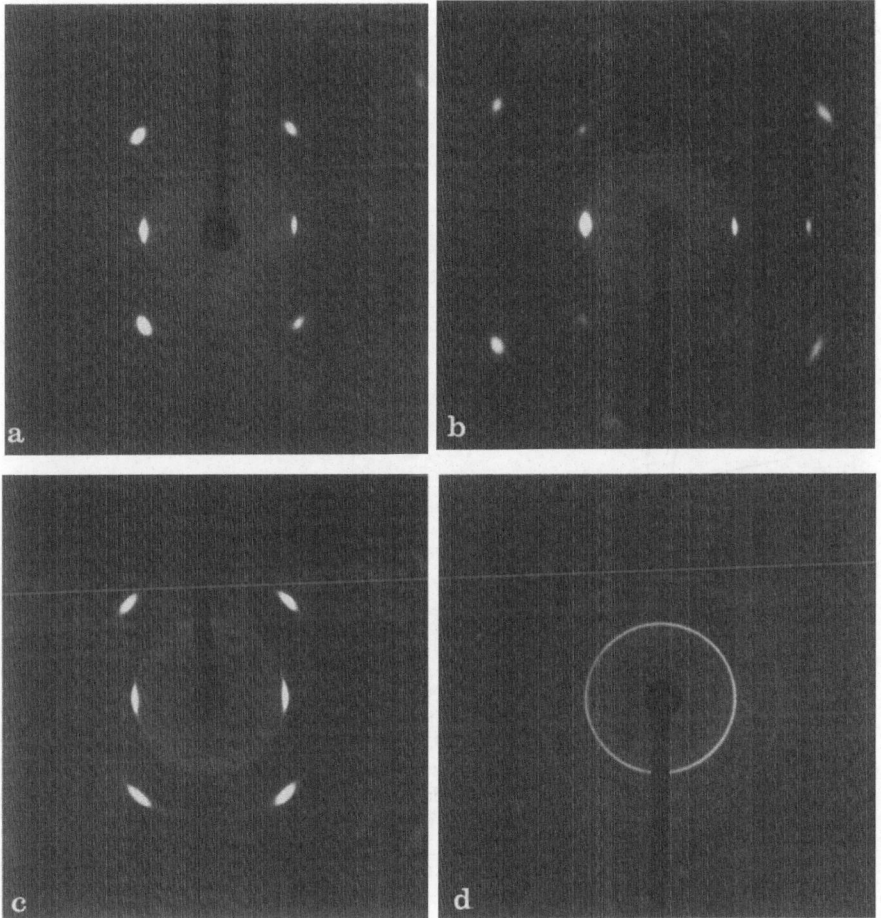

Fig. 4.9 a–d. Wide-angle X-ray scattering patterns of poly(tetroxocane) annealed in nitrogen: **a** original; and annealed, **b** at 160 °C, 30 min; **c** 181 °C, 2 min; **d** 195 °C, 1 min (melt-crystallized)

The change of the melting profiles by the annealing is represented by T_{s1} and T_{s2} as shown in Fig. 4.11, where the weight loss during annealing is less than 5% even at 181 °C. T_{s1} becomes higher in the annealing above 160 °C, while T_{s2} increases above 170 °C. The three plots of T_s below 170 °C in the figure are of the crystal formed from the partial melt on cooling, and the one plot at the annealing temperature of 195 °C is of the melt-crystallized sample.

4.5 100 Å Scattering

It is well-known that lamellar thickening occurs by annealing. Figure 4.12 shows the relationship between the annealing temperatures and the long spacings of 100 Å scattering. The long spacing of 100 Å scattering remains constant in the annealing below 140 °C, while it increases above 150 °C, and abruptly above 165 °C.

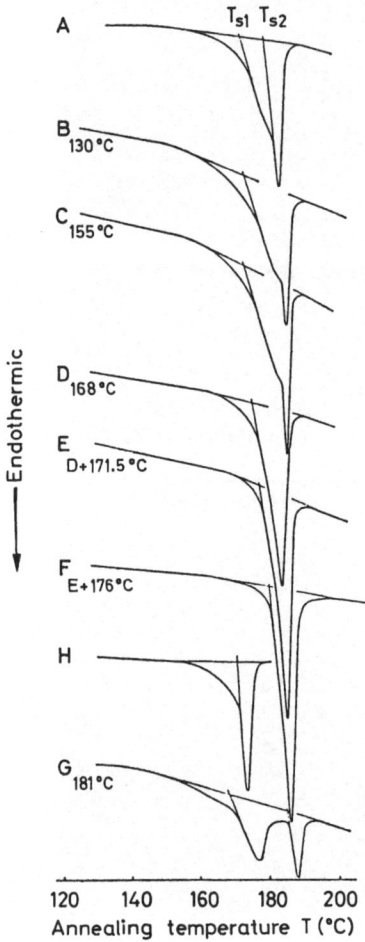

Fig. 4.10. Heating curves of poly(tetroxocane) annealed in nitrogen: Heating rate, 16 °C/min; A, original; and annealed, B, 130 °C, 30 min; C, 155 °C, 30 min; D, 168 °C, 10 min; E, sample D annealed at 171.5 °C for 15 min; F, sample E annealed at 176 °C for 12 min; G, 181 °C, 2 min; H, 195 °C, 1 min (melt-crystallized)

The long spacing of the melt-crystallized form is about 190 Å as indicated by filled circles in Fig. 4.12.

In the samples annealed directly at 175° and 181 °C two intensity maxima are observed. One long spacing at a higher angle corresponds to that of the melt-crystallized form, 190 Å, the other at a lower angle to the developed 100 Å scattering. But the sample annealed below T_s of the starting material gives only one spacing of the developed 100 Å scattering.

The increase of the 100 Å scattering occurs in the annealing above 150 °C as shown in Fig. 4.12, which temperature is about 10 °C lower than the temperature from which T_s increases by annealing as shown in Fig. 4.11. That is, the increase of the 100 Å scattering takes place before the occurrence of the rearrangement of the polymer chains, indicating a very small enthalpy change in the annealing below 160 °C. However, the fine structure with the 100 Å scattering must be related to the structure giving the melting endotherm at the lower temperature in the heating curve, because the increase of T_{s1} corresponds to the increase of the long spacing.

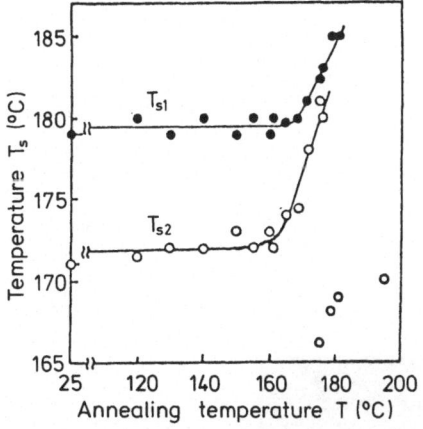
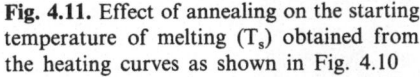

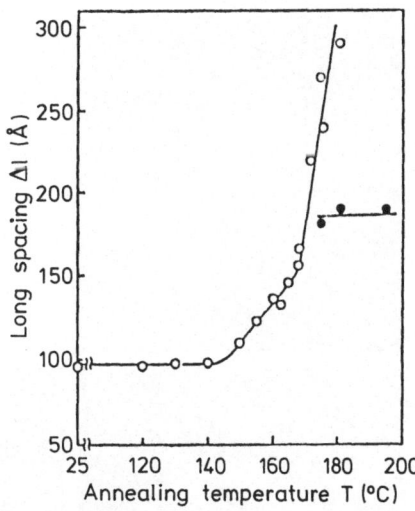

Fig. 4.11. Effect of annealing on the starting temperature of melting (T_s) obtained from the heating curves as shown in Fig. 4.10

Fig. 4.12. Relationship between annealing temperatures and long spacings of about 100 Å in an original sample ($\bigcirc$), and the melt-crystallized sample ($\bullet$)

4.6 Melt-crystallized Poly(oxymethylene)

The thickening of a lamellar crystal by annealing was studied in the samples of melt-crystallized POM. Figure 4.13 shows the SAXS patterns of melt-crystallized (a) and annealed samples (b ~ e). The ring-like scattering of the original sample becomes smaller in diameter upon annealing, indicating an increase of the lamellar thickness. The WAXS patterns, though the figures are eliminated, also show ring-like scattering indicating no preferential orientation of the polymer crystal.

Figure 4.14a shows the relationship between the long spacings of the melt-crystallized samples and the annealing temperatures. The long spacing of the original sample increases with annealing temperatures above 150 °C, in a similar manner to the cases of POM or its copolymer [6].

Figure 4.14b shows that T_s of the original sample shifts to a higher temperature in the annealing above 150 °C. It is, therefore, obvious that when the long spacing increases by annealing, T_s of the annealed sample also increases.

The rearrangement of the polymer chains of the melt-crystallized form in the annealing causes the increase of both long spacing and melting point.

In conclusion, the thickening of the lamellar crystal occurs if the annealing is performed at a temperature above 150 °C but below T_s of the starting sample.

4.7 Comparison of the Annealing Effect of two Kinds of Crystallites Mentioned in 4.5 and 4.6 [13a]

Figure 4.15 shows the relative increases of the 100 Å spacing of PT_EOX and that of the long spacing of the melt-crystallized POM. The increments Δl of the long spacing from the original in both samples indicate similar thickening of lamellae in

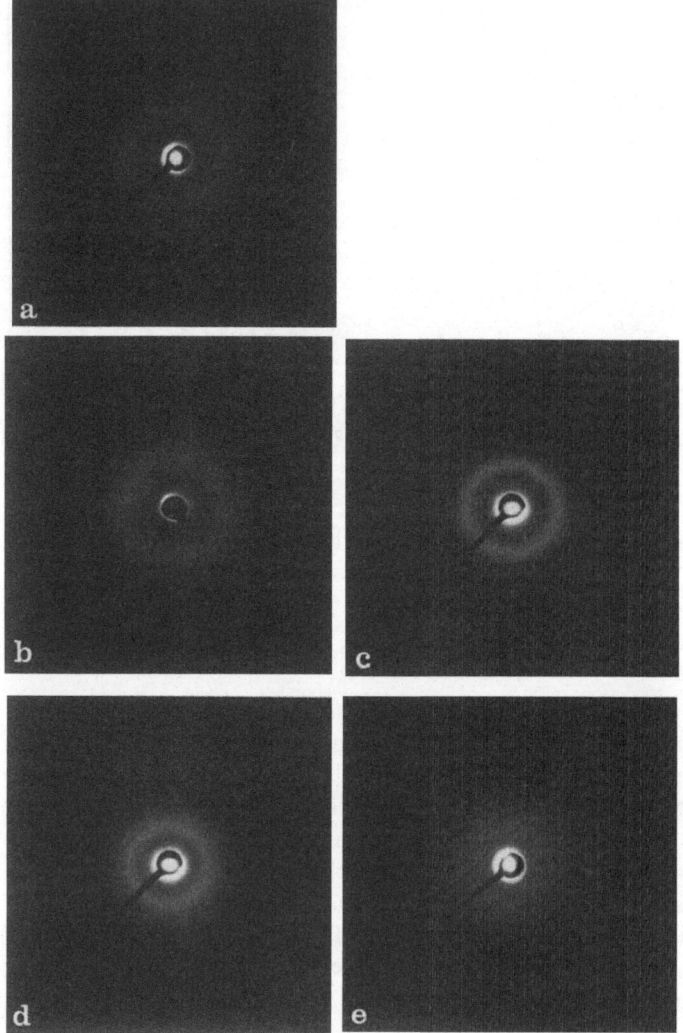

Fig. 4.13a–e. Small-angle X-ray scattering patterns of the melt-crystallized poly(oxymethylene), the acetylated poly(trioxane): **a** original (melt-crystallized); and annealed, **b** at 156 °C for 30 min; **c** 165 °C, 20 min; **d** 168 °C, 20 min; **e** sample **d** annealed at 172 °C for 10 min. Intensities are not comparable

the annealing above 150 °C. Similar results were obtained from the reported values in the references [25].

Figure 4.16 shows the relationship between T_{s1} and the reciprocal of the long spacing in both samples. The lines were evaluated by the least squares method. Provided T_{s1} behaves as T_m, it is related to the long spacing l by:

$$T_m = T_m^{\circ} (1 - 2\sigma_e/1 \cdot \Delta H_m)$$

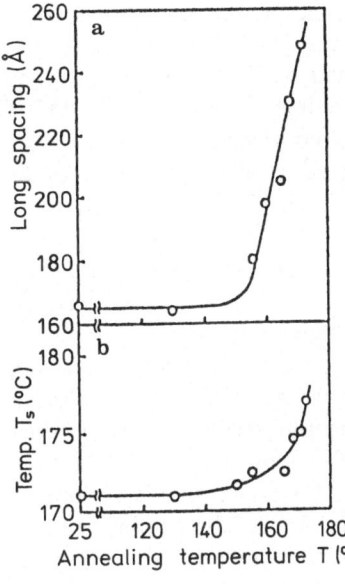

Fig. 4.14a and b. Relationship between annealing temperatures and long spacings of the melt-crystallized poly(oxymethylene) (**a**), and starting temperatures T_s of the melting (**b**)

where T_m is the melting point of the sample, $T_m{}^\circ$ the equilibrium melting point, σ_e the end-surface free energy, and ΔH_m the enthalpy of melting [26].

The linear relation between T_s and $1/l$ observed. The extrapolated peak temperature of an endotherm to the sample weight zero for the melt-crystallized sample has been applied to the above equation replacing T_s, which leads to a quite similar result.

The similarity of the equilibrium melting temperature, at 185 °C, is observed in both samples; this fact indicates a similar packing of the polymer chains between the 100 Å spacings in PT_EOX and lamellae in the melt-crystallized POM.

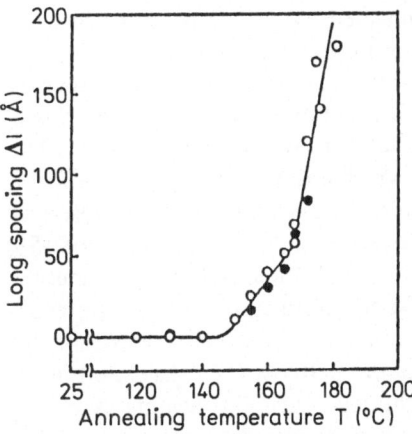

Fig. 4.15. Relative increase of long spacing (Δl) from the original value in the annealing: ○, poly(tetroxocane); and ●, melt-crystallized poly-(oxymethylene)

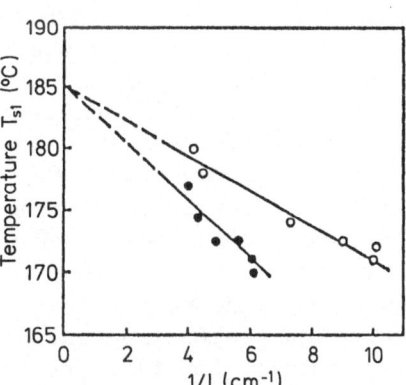

Fig. 4.16. Relationship between starting temperature of the melting at the lower temperature side (T_{sl}) and the reciprocal of the long spacing: ○, poly(tetroxocane); ●, melt-crystallized poly-(oxymethylene)

The melting point must be equivalent in both samples at the extreme condition (infinite lamellar length), provided the native structure is of the same type, since the melting point (T_m) is thermodynamically defined as $T_m = \Delta H_m / \Delta S_m$, where ΔH_m and ΔS_m are the enthalpy and the entropy of melting, respectively. The 185 °C equilibrium melting temperature is rather similar to the reported values [27].

In conclusion, the fine texture having a long spacing of about 100 Å in the as-polymerized PT_EOX without the sub-crystal is of lamellar type, since it shows a quite similar annealing effect with that of the lamellar crystal in the melt-crystallized sample. However, the slope of the straight line in Fig. 4.16 is greater in the melt-crystallized form than in the as-polymerized PT_EOX. The slope represents the ratio of the end-surface free energy (σ_e) of the lamellae to the enthalpy of melting (ΔH_m). The difference of the slopes suggests a difference of the lamellar stacking in the sample; the lamellar crystallites in PT_EOX are situated between fibrillar bundles of the extended chain crystal with a preferential orientation, while those in the melt-crystallized form aggregate in the form of spherulites.

The following data may support our suggestion: (1) Electron micrographs of PT_EOX show fibrils with ripples as shown in Section 5; (2) The heating curve of PT_EOX shows an inseparable double endotherm, which is quite different from that of the sample annealed at 181 °C; (3) The rearrangement of the polymer chains takes place in annealing even above T_{s1}.

4.8 500 Å Scattering [13b]

Figure 4.17 shows the SAXS patterns of as-polymerized PT_EOX, annealed for 30 min at various temperatures, and annealed for 8 days at 75 °C. The annealing at 75 °C for 100 h (about 4 days) brought no change in the SAXS patterns, but after 200 h a change was observed in the scattering patterns, as well as after 30 min in the temperature range between 100 °C and 140 °C. The broad scattering especially in the meridional direction in the SAXS patterns of annealed PT_EOX seems to be ring-like around the beam trap, suggesting the existence of unoriented voids.

The melting behaviors of the sample are shown in Fig. 4.18. The endothermic profile of the original PT_EOX changes slightly so that the two peaks become clearer by annealing, but the total area under the double peak remains constant. This fact indicates that the aggregation of the polymer chains are only minorly affected during annealing. Weight loss of the sample during annealing has been observed, as well as increases by raising the annealing temperature. A maximum loss of 15% occurs in the sample annealed at 140 °C. If such a weight loss is due to the thermal decomposition of high-molecular weight chains, the profiles of the melting endotherm as well as the X-ray scattering behavior must be influenced, seriously. No serious change, however, is noticed in the scattering patterns (Fig. 4.17) and in the melting behaviors (Fig. 4.18). Thus, the weight loss during annealing is caused by the sublimation of unreacted monomers and/or by the decomposition of low molecular weight chains.

The voids formed during polymerization retain unreacted monomers and/or low molecular weight polymers, which are easily removed during annealing because of the high vapor pressure of the former and the thermal instability of the latter

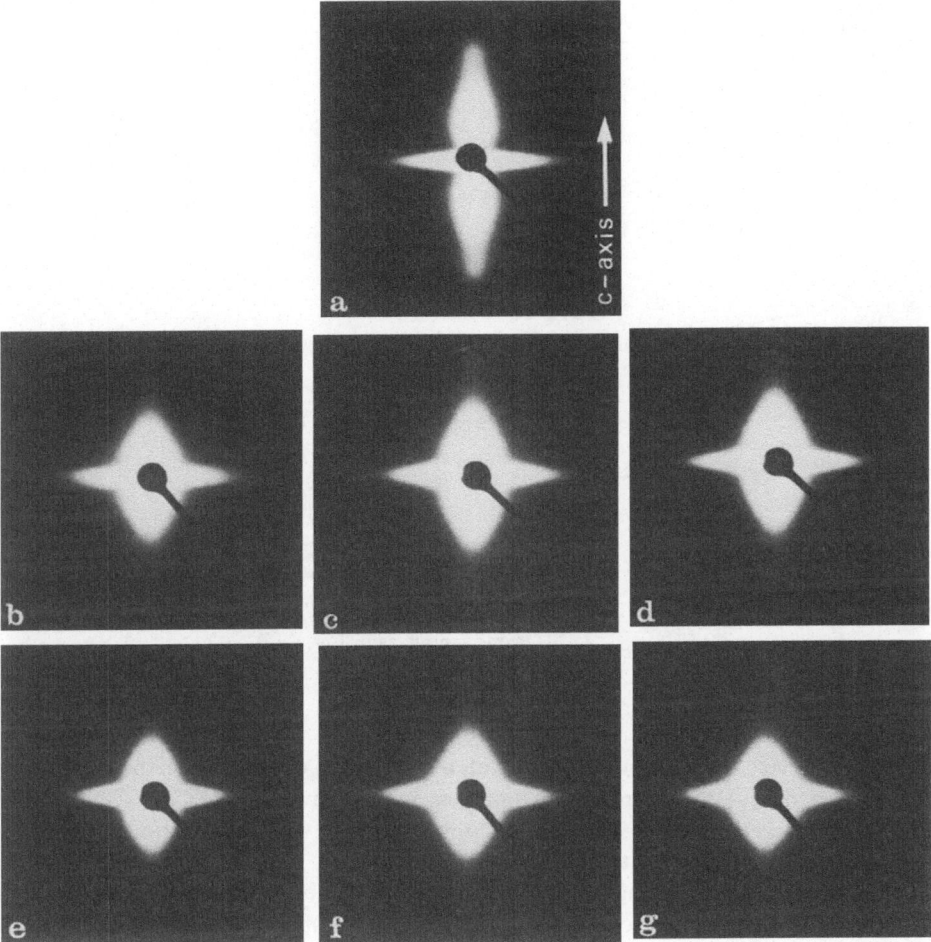

Fig. 4.17a–g. Small-angle X-ray scattering patterns of poly(tetroxocane) annealed at various temperatures for 30 min except at 75 °C for 8 days. The annealing temperatures are **a** Original; **b** 75 °C; **c** 100 °C; **d** 110 °C; **e** 120 °C; **f** 130 °C; **g** 140 °C. The intensities are not comparable

at elevated temperatures. These findings suggest that the fibrillar crystal, or the extended chain crystal, must be composed of the stacking crystallites.

Figure 4.19 shows the meridional SAXS intensity curves for PT_EOX post-polymerized in air. The original sample shows a very broad intensity maximum at about 13' of 2θ, the position of which shifts to the lower angle side by annealing the sample at a higher temperature. This indicates an increase in the spacing of the void appearance by the annealing.

Figure 4.20 shows the SAXS curves of PT_EOX post-polymerized at 105 °C in vacuum. The profile of the intensity curve of the original sample changes similarly to that of the sample obtained in air during annealing, indicating no special effect of oxygen on the 500 Å scattering. The length of the crystallites mentioned above

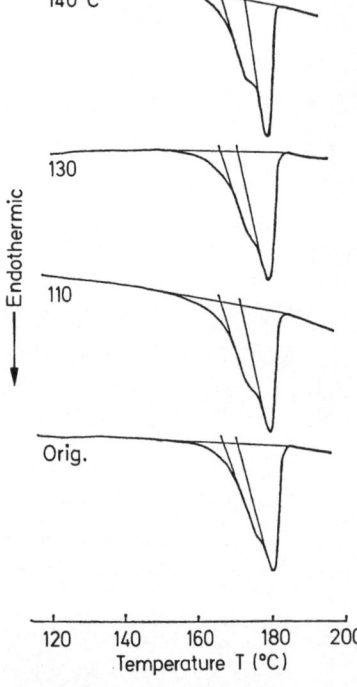

Fig. 4.18. Heating curves of annealed poly(tetroxocane) (same sample as in Fig. 4.17)

does not reflect the molecular weight of the polymer chains, since the post-polymerization of T_EOX in air gives a polymer with a lower molecular weight chain than that in vacuum [21,23]. The periodic appearance of voids on the polymer fibrilis obtained both in air and in vacuum is a characteristic of PT_EOX texture, relating mainly to the difference in unit cell dimensions of both crystals of monomeric T_EOX and PT_EOX. The broad SAXS intensity maximum in the meridional direction

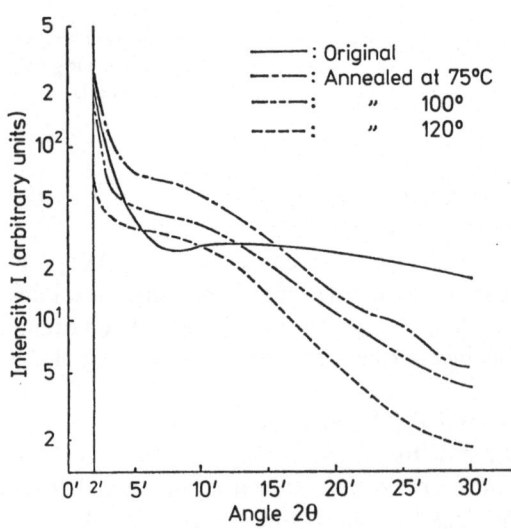

Fig. 4.19. Small-angle X-ray scattering curves of annealed poly(tetroxocane) (prepared in air). Annealing temperatures are indicated in the figure

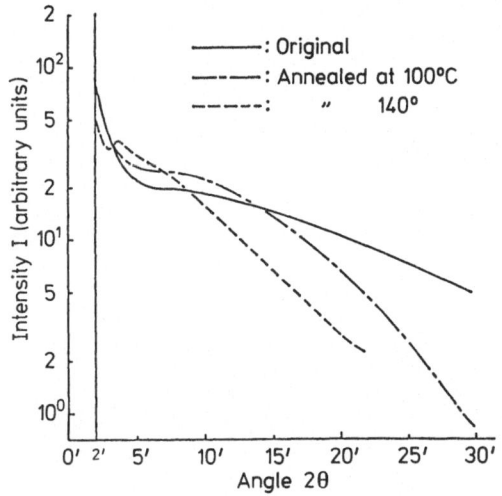

Fig. 4.20. Small-angle X-ray scattering curves of annealed poly(tetroxocane) (prepared in vacuum). Annealing temperatures are indicated in the figure

is caused by the materials in the voids. Annealing reduces these materials by the sublimation of unreacted monomer and/or by the thermal decomposition of low molecular weight polymers to emphasize the periodic appearance of voids.

4.9 Crystallite Size and Lattice Distortion

Here, annealing effects on the lattice constant, the dimension of the crystal, and the distortion parameter, are examined. Figure 4.21 shows the 001 reflection profiles of as-polymerized PT_EOX, annealed, and melt-crystallized POM (Delrin). By annealing the sample at 180 °C for 10 and 40 min the intensity peak shifts to a lower angles and obtains a similar profile as that of Delrin.

Figure 4.22 shows the 009 reflection profiles of PT_EOX as-polymerized at 105 °C and annealed. Annealing of the sample leads to the same behavior as in the case of PTOX. A slight shifted peak also was observed in the PT_EOX annealed even at 135 °C for 40 min. The 009 and 0018 reflections of PT_EOX were obtained at 80 °C (this polymer having no lamellar crystal as mentioned above) and annealed at 160 °C for 30 min.

In the same manner as in the case of PT_EOX obtained at 105 °C, the intensity peak of the annealed sample shifts to a lower angle and becomes symmetrical.

The changes of the lattice constant c and of half widths of 009 and 0018 profiles are calculated from the middle point of the half width of profiles by Bragg's condition. Table 4.1 summarizes the values of c. Besides the results of PT_EOX and PTOX, the values of poly(oxymethylene) (Delrin-500) are tabulated for comparison. All samples show the increase of c at an annealing temperature above 150 °C. PTOX annealed at 180 °C contains the folded chain crystal, since the heating curve shows a new endotherm at a lower temperature side, and Delrin is also of the folded chain crystal. In the case of the folded chain crystal, the increase of the lattice constant c is rather remarkable. On the other hand, the small increase of c in as-polymerized

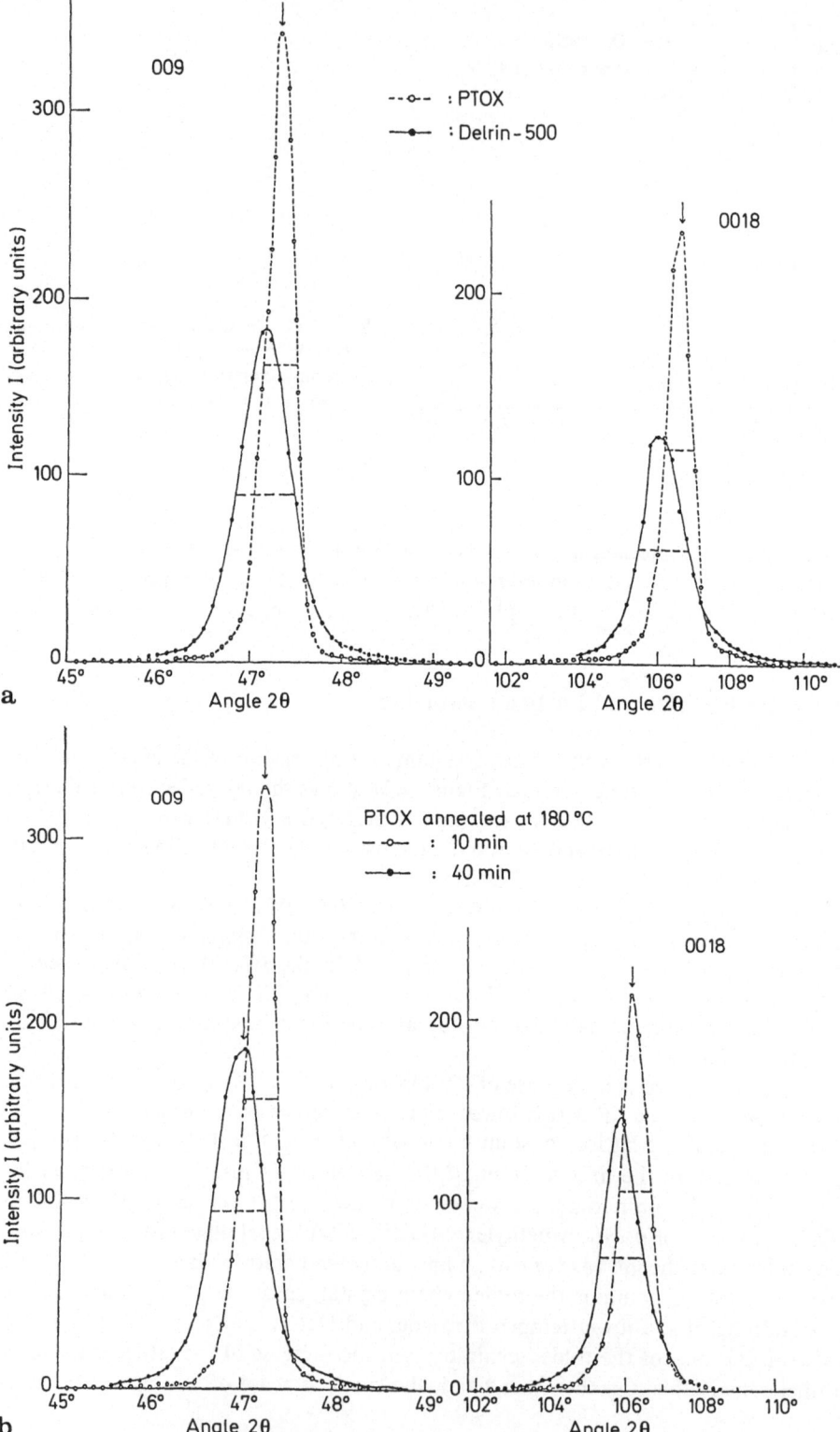

Table 4.1. Lattice constants c (Å) of poly(tetroxocane) obtained at 80 °C (PT$_E$OX-80) and at 105 °C (PT$_E$OX-105), poly(trioxane) (PTOX), and Delrin-500

Sample	Annealing condition		Lattice constant (Å)	
	Temp. (°C)	Time (min)	009	0018
PT$_E$OX-80				
As-polymerized	—	—	17.32	17.33
Annealed	135	40	17.34	17.35
	160	30	17.34	17.36
PT$_E$OX-105				
As-polymerized	—	—	17.37	17.37
Annealed	135	40	17.37	17.37
	170	120	17.38	17.38
	175	30	17.38	17.39
PTOX				
As-polymerized	—	—	17.29	17.31
Annealed	173	60	17.34	17.34
	180	10	17.34	17.34
	180	40	17.42	17.41
Delrin-500				
As-polymerized	—	—	17.32	17.33
Annealed	180	120	17.43	17.42

PTOX or PT$_E$OX may be due to the relaxation of the internal strain kept during the polymerization in the solid state as mentioned above.

The weight-average crystallite size $\bar{D}_w$, and the maximum distortion parameter g are obtained by line broadening analysis. The results are summarized in Table 4.2. In PT$_E$OX obtained at 105 °C, g decreases by annealing at 170 °C for 2 hours. In PTOX no special change is observed except when annealed at 180 °C for 30 min. In this case, the folded chain crystal is formed inbetween the extended chain crystals, or main crystals, to give larger g values.

The assumption mentioned above can be confirmed at least in the case of PT$_E$OX. These findings suggest that the extended chain crystals may be longitudinally compressed by the stress field due to the monomer matrix in the polymerization process.

5 Radiolysis

The effect of irradiation on the fine structure of the polymer is investigated in order to clarify the original morphology.

◀ **Fig. 4.21.** Wide-angle X-ray scattering profiles of 009 and 0018 reflection of poly(trioxane), Delrin a and annealed at 180 °C for 10 and 40 min. b.

Table 4.2. The weight-average crystallite size ($\bar{D}_w$) and the maximum distortion parameter (g) of poly(tetroxocane) (PT_EOX-105) and poly(trioxane) (PTOX)

Sample	Annealing condition		Results	
	Temp. (°C)	Time (min)	$\bar{D}_w$ (Å)	g (%)
PT_EOX-105				
As-polymerized	—	—	550	1.6
			100	0.9
Annealed	170	120	250	1.2
PTOX (main-crystallite)				
As-polymerized	—	—	550	0.7
Annealed	173	60	450	0.6
	180	10	400	0.7
	180	30	250	1.5
PTOX (sub-crystallite)				
As-polymerized	—	—	250	0.7
Annealed	173	60	300	0.8
	180	10	250	0.8

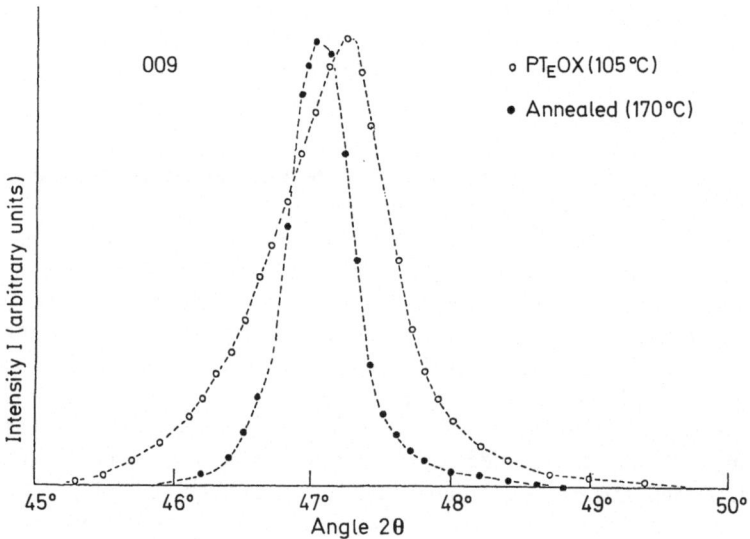

Fig. 4.22. Wide-angle X-ray scattering profiles of 009 reflections of poly(tetroxocane) prepared at 105 °C, as-polymerized, and annealed

5.1 Melting Behavior of Irradiated PTOX

A double endotherm in the heating curve of PTOX is obtained irrespective of the polymer yield, but the profile changes with the heating rate in DSC or DTA measurements [14a].

Figure 5.1 shows the heating curves of PTOX [6% (a) and 80% (b) polymer yield] irradiated with various doses on an open system. The profiles of a 6% yield are not affected by irradiation even after 1 kGy. The profiles of an 80% yield, however, changes dramatically with irradiation, that is, the samples irradiated below 0.1 kGy give similar heating curves to the original PTOX, but those irradiated above 0.25 kGy show that the endotherm at a higher temperature side (higher endotherm) almost disappears. The samples irradiated over 1 kGy give a very sharp endotherm. The profiles of the endotherm irradiated over 1 kGy are similar to each other. The starting temperature T_s of the samples remains constant irrespective of dose, except for the sample irradiated at 10 kGy.

The SAXS and WAXS patterns of the sample irradiated up to 10 kGy are identical to those of the original sample. The SAXS pattern of the original shows a diffuse scattering in the equatorial direction, but little scattering meridionally and the WAXS patterns show reflection caused by sub-crystals, as well as patterns caused by main crystals (fibrous reflections) as reported by Chatani et al. [8]. So only the melting behavior is affected by irradiations below 10 kGy as shown in Fig. 5.1 b.

Figure 5.2 shows the thermogravimetric (TG) curves of the sample irradiated up to 10 kGy (the same sample as used in Fig. 5.1). The weight loss due to the thermal decomposition seems to occur in two stages; the first stage in the TG curves is observed at about 200 °C, the second at about 300 °C. Irradiation reduces the decomposition during the first stage and shifts the starting temperature of the decomposition to the lower temperature side.

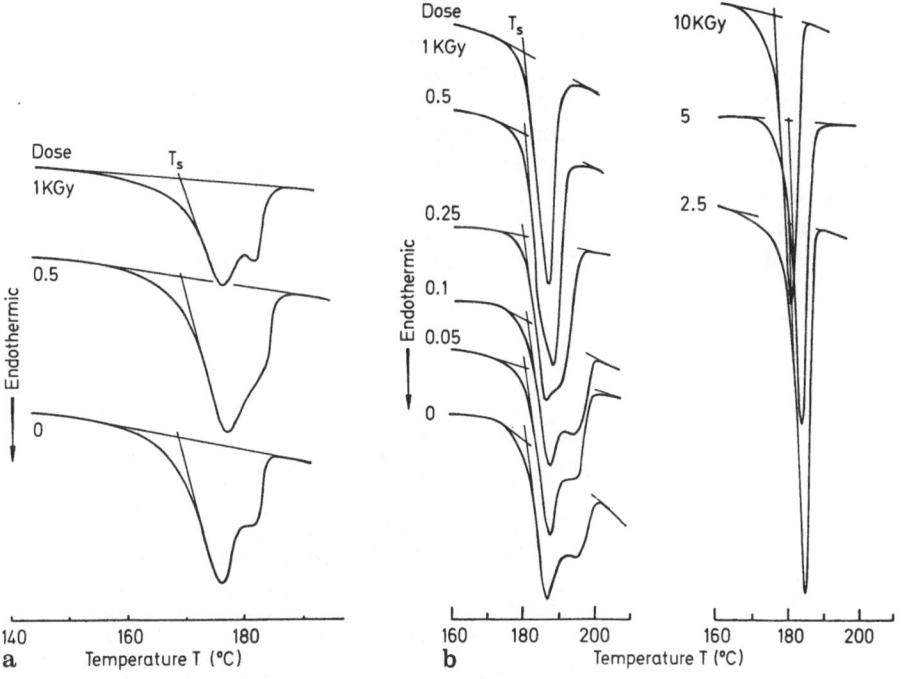

Fig. 5.1 a and b. Heating curves of poly(trioxane) irradiated with various doses: **a** 6% polymer yield; **b** 80% polymer yield; heating rate, 160 °C/min. The exposure dose is indicated in the figure

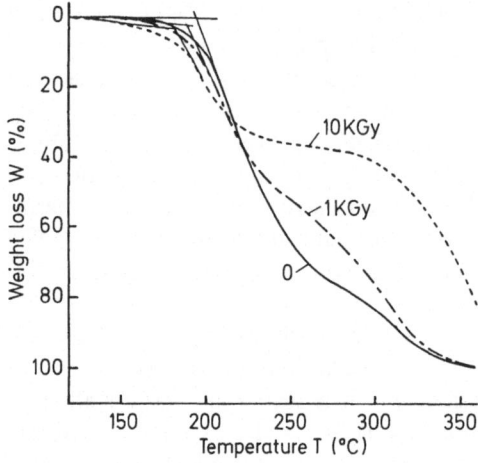

Fig. 5.2. Thermogravimetric curve of poly-(trioxane) irradiated with various doses: Heating rate, 10 °C/min. The exposure dose is indicated in the figure

Figure 5.3 shows the heating and TG curves of the powdery PTOX rolled with shear stress. The sharp endotherm in the heating curve due to the melting is a singlet, thus the higher temperature endotherm shown in Fig. 5.1 is lost. A broad endothermic peak at about 280 °C is due to the decomposition occurring monotonically instead of in two stages as in the original sample (unirradiated) in Fig. 5.2.

The amount of polymer decomposed during the first stage of the TG curve is reduced by irradiation, indicating that chain scission occurs, since a thermally stable polymer is produced after irradiation [28].

A reduction of the higher endotherm in the heating curve by irradiation is associated with chain scission. A similar situation is expected in PTOX rolled by a mixing mill, i.e., chain scission occurs through shear stress.

Figure 5.4 shows the heating curves of melt-crystallized PTOX cooled from the melt at a rate of 64 °C/min, or the 2nd run of the 80% sample used in Fig. 5.1. The profile of the heating curve changes more sharply with increasing dose, although T_s remains constant in all samples, except those irradiated with 10 kGy. Quite similar results were obtained in two types of POM, a spherulite crystal and an oriented

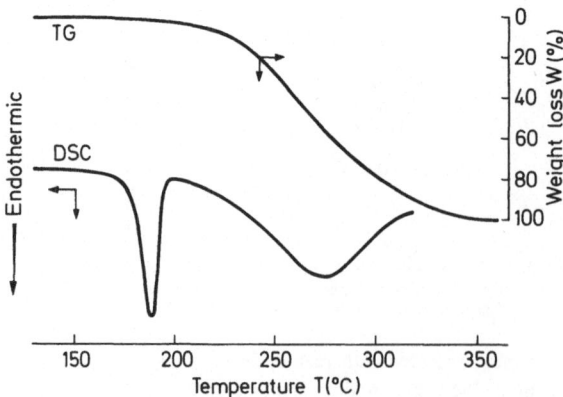

Fig. 5.3. Thermogravimetric curve and heating curve of poly(trioxane) (80% yield) transformed into powder by mixing mill: Heating rate, 10 °C/min in TG, and 16 °C/min in DSC heating curve

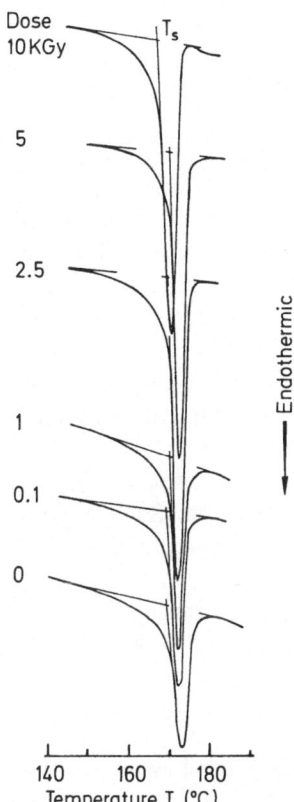

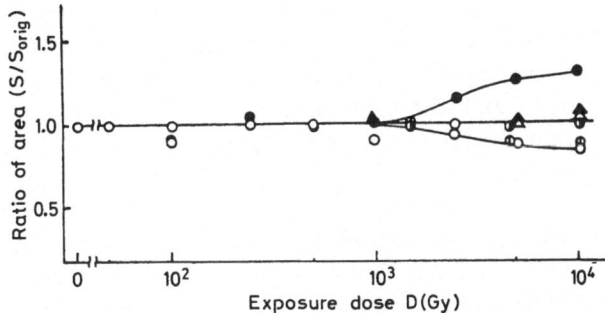

Fig. 5.5. Relationship between the ratio (S/S$_{orig}$) of the area per unit weight under the endothermic peak of the irradiated (S) to the unirradiated (S$_{orig}$) and dose (D): ○, poly(trioxane); ◑, Delrin; △, Takafest, acetylated poly(trioxane)

Fig. 5.4. Heating curves of the melt-crystallized poly(trioxane): Heating rate, 16 °C/min. The values in the figure indicate the exposure dose

lamellar sample (the drawn film). No remarkable difference was found between the unirradiated and the irradiated up to 10 kGy.

Figure 5.5 shows the effect of irradiation of three types of samples upon the area under the endothermic peak in the heating curves, which corresponds to the enthalpy of melting.

It is remarkable that the area in the 2nd run of PTOX increases with irradiations over 1 kGy compared with the original, indicating an increase in enthalpy of melting. The area of the other samples, i.e., the oriented lamellae and the spherulite crystals, after irradiation remains the same or a little less than that of the unirradiated, indicating no special change in the enthalpy of melting. This fact may indicate that chain scission occurs probably at a non-crystalline region to increase only the uniformity of the melting behavior of the polymer crystal.

On the other hand, although chain scission by irradiation may occur in as-polymerized PTOX as well, the profile of the melting endotherm and the area under the endothermic peak show different behavior in the irradiated PTOX, indicating a difference in aggregation of polymer molecules or morphology.

It is suggested that the higher endotherm may be due to the superheatability of the PTOX crystal, which is reduced by chain scission, most likely on polymer chains at a particular region of entropy restriction.

The following facts support this suggestion; the superheatability of polyethylene with a cross-banded micro shish kebab structure [29] was reduced by fuming nitric acid treatment. This reduction is associated with only a limited amount of molecular scission and with the scission of the interplatelet tie molecules.

Superheatability of poly(ethylene terephthalate) was produced, or in some cases restrained in the sample annealed at about 250 °C, which was not observable for the as-polymerized sample [30]. This was not an effect of crystal size, but must be caused by tie molecules between different crystals or different locations on the same lamellae. Superheating effects due to reduction in entropy of melting are calculated theoretically [31] in a situation where each chain molecule can simultaneously belong to many different crystalline and noncrystalline regions. In the case of natural rubber, the chain scission of molecules with stress, i.e., a large deformation with an extention ratio of more than 2, occurs easily by irradiation, since the molecules are entropy restricted [32].

In conclusion, it is suggested that chain scission takes place at a selective region in PTOX where molecular chains are probably entropy restricted. Such regions are superheatable. Irradiation over 250 Gy, or shear stress, must reduce these selective regions.

The effect of irradiation over 10 kGy on the melting and X-ray scartering behaviors will elucidate more clearly the chain scission at a selective region in PTOX.

5.2 Layer-Like Voids in Irradiated PTOX

PTOX is irradiated over 10 kGy, and the SAXS and WAXS behaviors are investigated in detail [14b].

Since the melting behavior changes rather monotonically by irradiation, the endotherm of the irradiated samples becomes very broad and shifts to a lower temperature side. Irradiation products, moreover, are found in the sample, which are easily evapolated or decomposed during heating. These phenomena complicate the melting behavior.

Figure 5.6 shows the SAXS patterns for PTOX as-polymerized (a), and irradiated with various doses up to 1 MGy (b, c, and d). The long axis of a bundle-like specimen, c-axis, is parallel to the meridian of the SAXS patterns.

Diffuse scatterings are more extended along the equatorial direction than along the meridional. The diffuse scattering along the meridian in as-polymerized PTOX is very weak, even in the vicinity of the primary beam, and no discrete scattering is observed. Over 0.3 MGy, however, a striated reflection can be seen on the meridian. These patterns of the irradiated sample shows that the maximum of the discrete scattering shifts toward a higher angle, and also the shape of the reflection, i.e., the length along the layer line and the width of the reflection, changes with increasing doses. Concurrent changes in the scattering intensity distribution are investigated with a Kratky U-slit collimation. Figure 5.7 shows how the meridional diffraction peak shifts to a higher angle and increases in intensity as the exposure dose increases, and how the meridional diffuse (continuous) scattering increases.

In contrast with γ-irradiation, the heat treatment at 179 °C in air gave only a little increase in scattering intensity on the meridian [33], and the sample treated in nitrogen gas did not give the intensity peak associated with the long period [9,33].

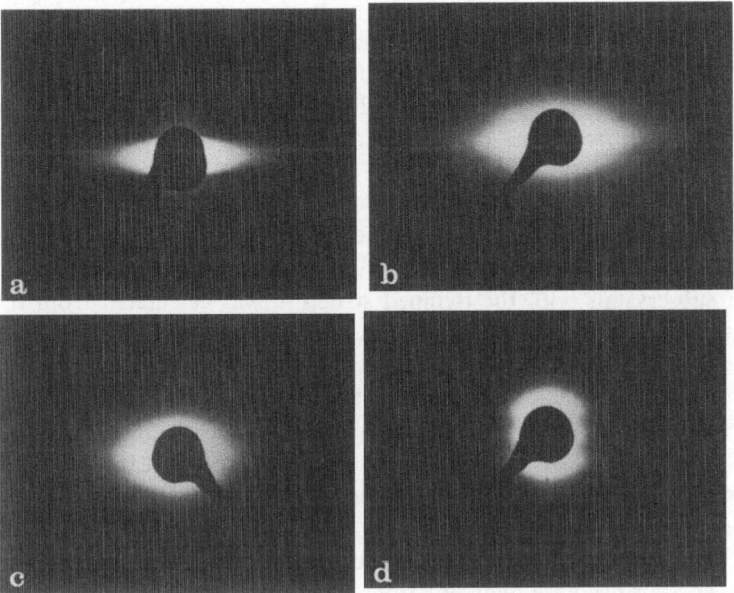

Fig. 5.6a–d. Small-angle X-ray scattering patterns of poly(trioxane) irradiated with various doses: **a** original (unirradiated); and irradiated, **b** 0.3 MGy; **c** 0.5 MGy; **d** 1.0 MGy. Intensities are not comparable

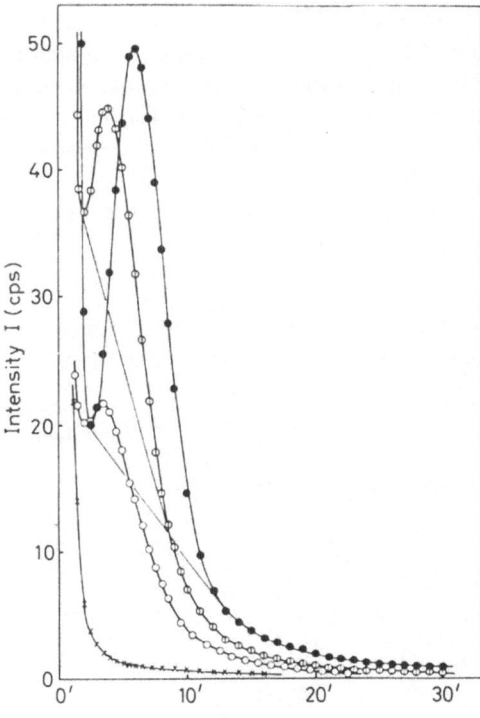

Fig. 5.7. Effect of irradiation on the small-angle X-ray scattering intensity in the meridional direction of poly(trioxane) using a Kratky U-slit camera: ×, original; irradiated, ○, 0.3 MGy; ⦵, 0.5 MGy; ●, 1.0 MGy

In the WAXS pattern of the irradiated sample, no trace of an amorphous halo is detected inside the 100 reflection peak. As the irradiation dose increases, spots due to crystal reflections fade, but show no change in the orientation of micro-fibrils. The intensity of twin-reflections from the sub-crystal decreases more rapidly than those from the main crystal with increasing irradiation dose, but the twin peak does not disappear completely.

The 100 and the 005 reflection peaks were observed in the WAXS diagram, indicating the existence of large amounts of unoriented micro-crystals and conformational disorder, as mentioned in Section 2.

The fraction of sub-crystals with the twinned structure was calculated from the intensity ratio of the main- and sub-100 reflections after corrections for the Lorentz factor were made, and also the amorphous content and the amounts of randomly oriented crystallites were calculated as shown in Table 5.1. The fraction of the sub-crystal in the unirradiated sample of the 80% polymer was 40%, which is equivalent to the value (45%) reported by Carazzolo et al. who did not report on the yields [34]. The fractional amount of the sub-crystals decreases with increasing polymer yield as mentioned in Section 2.

All of the fractional amounts of main and sub-crystals, amorphous content, and randomly oriented crystals decrease by irradiation. The reductions of the sub-crystal and randomly oriented crystal are more extensive than the main. The amorphous content was reduced completely in PTOX irradiated over 0.3 MGy. A weight loss took place as a result of irradiation in air as also shown in Table 5.1, indicating the depression of the melting point and the broadening of the endothermic peak by vigorous radiolysis.

Figure 5.8 shows the electron micrographs of PTOX irradiated with various doses. The micro-fibrils of PTOX change into a mosaic-like stack of crystallites by irradiation, and the larger the dose the more uniform the observable stacking. These patterns as well as the SAXS patterns in Fig. 5.1 indicate the formation of a layer-like void perpendicular to the c-axis of fibrillar crystals with a period of about 1000 Å.

The analysis of the observed SAXS curves is performed with the following items.

Table 5.1. Reductions of fractional amounts of main crystals, sub-crystals, amorphous contents, and randomly oriented crystals, and weight loss of poly(trioxane) irradiated in air at room temperature

Exposure dose (MGy)	Reductions of fractional amounts (%)				Weight loss (%)
	main-crystal	sub-crystal	amorphous	randomly oriented crystal	
0 (as-polymerized)	0 (initially 51%)	0 (initially 40%)	0 (initially 4%)	0 (initially 5%)	0
0.3	0 (51)	20 (32)	100 (0)	60 (2)	20 ± 5
0.5	7 (47)	20 (32)	100 (0)	80 (1)	30 ± 5
1.0	25 (38)	25 (30)	100 (0)	100 (0)	50 ± 5

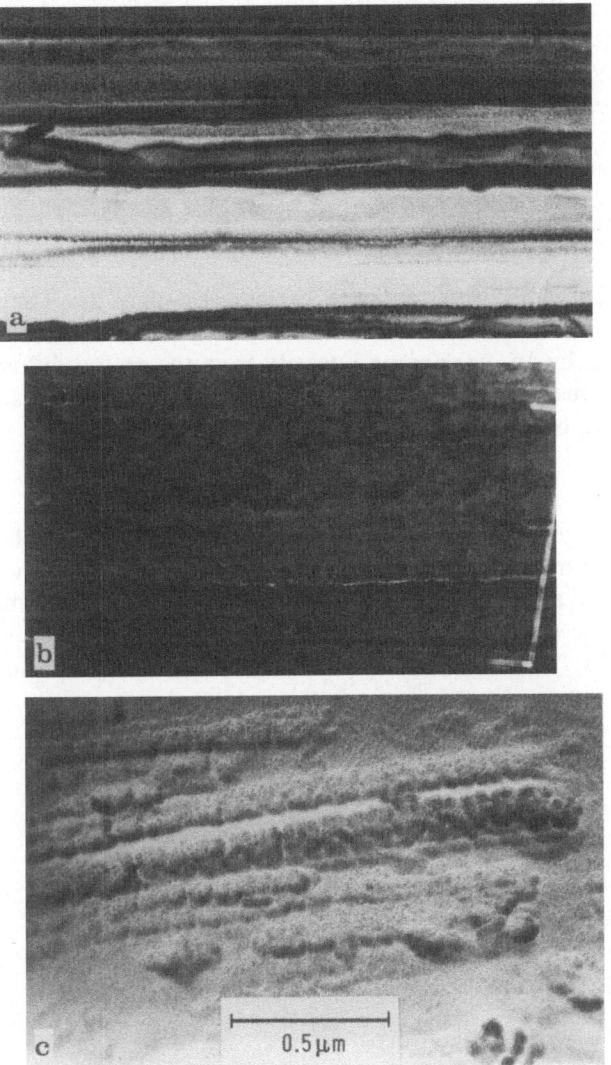

Fig. 5.8a–c. Electron micrographs **a** of the free surface of as-polymerized poly(trioxane) crystals; **b** irradiated with 0.5 MGy; **c** 1.0 MGy

1) Layer-like voids in micro-fibrils of the irradiated samples

The maxima in the SAXS patterns shown in Fig. 5.6 and 5.7 are characteristic of a super-lattice structure of one-dimentional heterogeneity in electron density and are regarded as evidence of the existence of a "long period". In drawn semi-crystalline polymers, this maximum is observed most distinctly on the meridian of the SAXS and is considered to be due to the alternating layers of dense crystalline and less dense amorphous regions. In as-polymerized PTOX the amorphous region is quite small. The present investigations of the irradiated PTOX also show no evidence of the amorphous halo detectable on the WAXS patterns as mentioned above. Therefore, it is suspected that the meridional layer reflection in the SAXS mentioned above is evidence of porosity caused by layer-like voids formed along the long axis of the fibrillar bundle. These voids may be associated with the weight loss of the irradiated polymer. The horizontal width of the meridional reflection can be used to evaluate the transverse dimensions of layer-like voids with the Scherrers' Equation:

$$\Delta\psi_{1/2} \simeq 0.9 \cdot \lambda/d \tag{1}$$

where $\Delta\psi_{1/2}$ is the angular half-width of the streak, λ the wave length of the X-ray, and d the void-width. As shown in Fig. 5.6, accurate measurements of the intensity along the streak are restricted by an overlapping intensity from the central diffuse scattering. It can be stated qualitatively, however, that the void-width increases from 200 Å to 300 Å as the exposure dose increases from 0.3 to 1 MGy.

2) Tsvankin's model

The distribution of crystallites along a micro-fibril does not appear to form a regular periodic system, because only a first-order maximum can be seen in the intensity curve. The first-order diffraction peaks allow the estimation of the Bragg spacing. The apparent long periods, L, thus calculated for the irradiated samples are given in Table 5.2. The true long period, D, can be calculated from the SAXS curve according

Table 5.2. Positions and widths of maxima on SAXS curves for PTOX irradiated with various doses

	Exposure Dose (MGy)		
	0.3	0.5	1.0
Observed Long Period - (Å)	1500	1300	900
Half Width $\theta_{1/2}(\times 10^{-4}\ \text{Å}^{-1})$	—	5.78	7.66
Width Parameter $q\ (= L \cdot \theta_{1/2})$	—	0.75	0.69

to Tsvankin's [35] and Blundell's [36] analyses. Recently, Crist [37] critically reviewed the various proposed models of a paracrystalline one-dimensional super-lattice.

In Tsvankin's theory, we assume the following model of the irradiated micro-fibril: (1) Zones of different density (crystal and void zones) form a super-lattice which is responsible for the long period; (2) The crystallite length c is uniformly distributed with the mean length $\bar{c}$, the dispersion having a mean value represented by σ_c; (3) The electron density as a function of distance in the fibre-axis direction Z is:

$$p(Z) = \begin{cases} = 0 \quad \text{for} \quad Z > c & (2\,a) \\ = \dfrac{1}{\bar{v}} \exp\left(-\dfrac{Z-c}{\bar{v}}\right) \quad \text{for} \quad Z \geq c & (2\,b) \end{cases}$$

where $\bar{v}$ is the mean value of the void length between the crystallites, so that $D = \bar{Z} = \bar{c} + \bar{v}$ represents the true periodicity of the super-lattice.

3) Analysis by Tsvankin's method

The background intensity of the diffuse scattering must be taken into account. If the electron density is distributed irregularly, the intensity of the diffuse scattering decreases monotonically with the scattering angle. Figure 5.7 shows that after irradiation the longitudinal distribution of density becomes less regular. It is not clear, however, whether or not the diffuse scattering and the diffraction peak arise in the same heterogeneous regions.

For simplicity, the diffraction maximum was separated from the background fairly arbitrarily, as shown in Fig. 5.7, and then the experimental curves were analyzed with respect to the position and widths of the maxima to compute the mean crystallite and void lengths.

The experimental half-widths of the maximum Δ_e were corrected by subtracting the influence of broadening of the primary beam as follows:

$$\theta_{1/2}^2 = \Delta_e^2 - \Delta_p^2 \tag{3}$$

where Δ_p is the primary beam half-width, 1.7 min in the present case.

The values of the corrected half-width of the maximum $\theta_{1/2}$ in unit of $2 \sin \theta/\lambda$ and the width parameter $q = L \cdot \theta_{1/2}$ for the irradiated PTOX are given in Table 5.2.

Difficulties with this simple approach are that apriori knowledge of the following parameters is required: the dispersion of crystallite lengths, σ_c; the crystallite packing density along the fibre axis (or the linear crystallinity), $\varrho_l = \bar{c}/d$; and the one-dimensional variation in electron density at each crystal — void boundary ε. Tsvankin [35] noted that the shape of the scattering curve was relatively insensitive to the value of ε chosen. The large values of the width parameter q, which are 0.75 and 0.69 for the samples irradiated with 0.5 and 1.0 MGy, respectively, as shown in Table 5.2, indicating a less regular periodicity. From these values of the parameter q, the true long period D, the mean crystallite length $\bar{c}$, and the linear crystallinity or the linear weight loss $\chi_l = 1 - \varrho_l$ can be estimated, once a suitable value is assigned to the parameter $\sigma_c/\bar{c}$ which characterizes the dispersion of crystallites around the mean value. When the broadness of the first maximum and the fact that no second order was observed are taken into account, the parameter $\sigma_c/\bar{c}$ can not be assumed to be small. According to Buchanan's results [38]. D, $\bar{c}$, and χ_l were obtained for the values of $\sigma_c/\bar{c}$ in the range of 0.2 to 0.5, when $\varepsilon = 0.2$, as shown in Table 5.3.

The loses of mass and sub-crystal content caused by irradiation enable us to estimate the value of $\sigma_c/\bar{c}$, assuming that the irradiation produces the electron density distribution described in Eq. 2.

For 0.5 MGy, $\sigma_c/\bar{c}$ probably takes a value between 0.4 and 0.5, since the weight loss is about 30%, as shown in Table 5.1. Thus the true long period of 1,200 ~ 1,300 Å and the crystallite length 800 ~ 1,000 Å could be determined. For 1 MGy, $\sigma_c/\bar{c}$ seems to be about 0.2, leading to the true long period of ca. 1,000 Å and the crystallite length of ca. 600 Å.

Table 5.3. Analysis of SAXS data for PTOX irradiated with 0.5 and 1.0 MGy

$\sigma_c/\bar{c}$	0.5 MGy			1.0 MGy		
	Long Period D (Å)	Crystallite Length $\bar{c}$ (Å)	Weight Loss χ_1 (%)	Long Period D (Å)	Crystallite Length $\bar{c}$ (Å)	Weight Loss χ_1 (%)
0.2	1560	790	50	1050	560	47
0.3	1310	690	48	960	530	45
0.4	1270	770	39	870	540	38
0.5	1170	1020	13	710	640	10

The mean crystallite length as well as lattice distortion is also obtained by X-ray line broadening analysis based on the paracrystalline theory.

The mean crystallite lengths in both parallel and perpendicular directions to the chain axis are estimated by using the intrinsic integral breadth of the 009 and 0018 reflections and of the four reflections from the 100 to 400.

Table 5.4 shows the values of the as-polymerized main and sub-crystallite of PTOX as well as values for the main crystallite irradiated with 1 MGy, since the WAXS intensity of the sub was very weak.

An averaged period of a stacking unit along the fibre axis can be the sum of the longitudinal length of the main (500 Å) and the lateral length of the sub-crystallite (400 Å), that is, about 900 Å. The decrease of the c-axis dimension of the main crystallite from 500 Å to 350 Å, and the decrease of the perpendicular dimension from 300 Å to 250 Å after irradiation with 1 MGy corresponds to a reduction of the weight loss of about 50%. Considering the bulk weight loss of ca. 50% shown in Table 5.1, the loss of the mass of the sub-crystallite must be in the same order.

The reductions in fractional amounts of the as-polymerized main and sub-crystallites during irradiation is in the order of 25% of the original main- and sub-crystallite from Table 5.1. The layer-like voids as well as local defects are formed by the irradiation of PTOX crystals, suggesting the periodical (ca. 1,000 Å) characteristics of the fibrillar crystal of PTOX. For example, the existence of a selective region accessible to radiolysis is assumed along the c-axis of PTOX crystals.

5.3 Melting and SAXS Behavior of Irradiated PT_EOX

Figure 5.9 shows the heating curves of PT_EOX irradiated with various doses. PT_EOX is prepared at 105 °C after preirradiation with 10 kGy, at an 80% yield.

Table 5.4. Crystallite length of PTOX parallel and perpendicular to the chain axis

	Parallel to c-axis (Å)	Perpendicular to c-axis (Å)
As-polymerized (Main)	500	300
(Sub)	200	400
Irradiated 1.0 MGy	350	250

Irradiation with 0.1 MGy mainly causes the depression of the melting point, but the profiles of the melting endotherm remains unchanged. An irradiation with more than 0.5 MGy changes the profile completely and a depression of the melting point occurs, indicating the introduction of a large amount of voids into the crystal. The SAXS intensity in the equatorial direction changes slightly by irradiation with up to 0.5 MGy, implicating that the size of voids between fibrils is not affected much by the irradiation though the melting behavior changes vigorously as shown in Fig. 5.9. The size of voids has been evaluated by electron microphotometry giving a length of 700 ~ 900 Å for the c-axis in both as-polymerized samples irradiated at 0.5 or 1.0 MGy.

Figure 5.10 shows the SAXS intensity curves of the same sample as in Fig. 5.9 (meridian). The intensity curves are normalized to the unit mass of the sample by correcting the weight loss during irradiation. The as-polymerized sample shows an intensity maximum at ca. 6', corresponding to a Bragg spacing of ca. 1000 Å, and a shoulder at a higher scattering angle. The position of the maximum is not affected by irradiation below 0.1 MGy, but affected greatly over 1.0 MGy.

The shoulder at a higher angle disappears after irradiation even at 4.5 kGy to show a single intensity peak.

These facts indicate that the long spacing of ca. 1000 Å in the as-polymerized sample becomes about 600 Å after 1.0 MGy, precisely after 0.5 MGy, the intensity of which remains constant per unit weight.

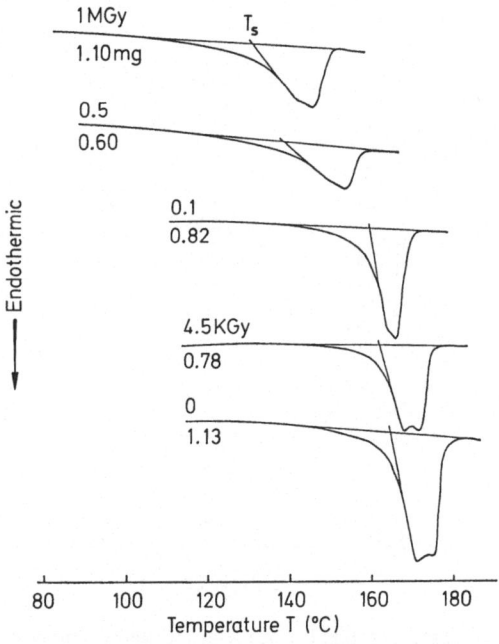

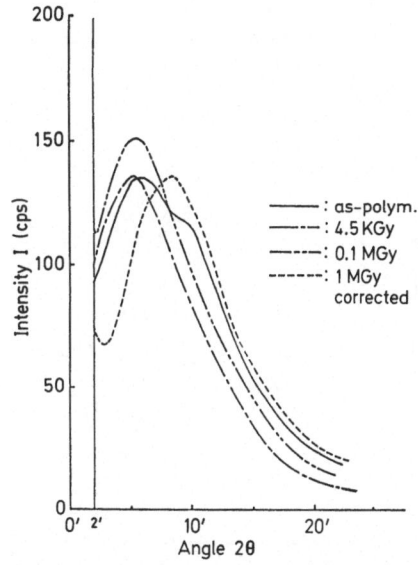

Fig. 5.9. Heating curves of poly(tetroxocane) irradiated with various doses: Heating rate, 16 °C/min. Doses and sample weights are indicated in the figure

Fig. 5.10. Small-angle X-ray scattering intensity curves of poly(tetroxocane) (obtained at 105 °C) irradiated with various doses: Curves are corrected to a unit weight of the sample

The formation of voids in the direction of the c-axis occurs by radiolysis with over 0.5 MGy causing the apparent contraction along the c-axis of the crystal. As mentioned in Section 2, PT_EOX has two different crystal lengths, at 500 Å and 100 Å, which are evaluated from the Hosemann plot with the 001 intensity in WAXS. Irradiation with more than 0.5 MGy renders the vigorous reduction of the reflection intensity corresponding to 100 Å unobservable, as well as the long spacing of 500 Å to ca. 200 Å. This fact indicates that the vigorous radiolysis takes place along the c-axis to reduce the crystal length as well as the long spacing, which is consistent with the result from SAXS.

The radiation effect is studied in PT_EOX obtained at a temperature below 90 °C, i.e., in PT_EOX containing sub-crystals along with main crystals.

Figure 5.11 shows the heating curves of PT_EOX obtained at 84 °C (14%, pre-irradiated with 10 kGy). The profile of the as-polymerized sample shows a single endothermic peak with a little shoulder, which remains almost constant even after irradiation with 0.1 MGy, though T_s shifts to a lower temperature side. Irradiation with over 0.5 MGy causes the melting endotherm in the heating curve to become very broad with lower T_s, indicating the formation of voids in the crystal in a similar manner to that observed in PT_EOX prepared at 105 °C.

The radiation effects on the melting behavior of the main and the sub-crystal or the lamellar crystal are similar. The SAXS behavior in the equator is also unaltered

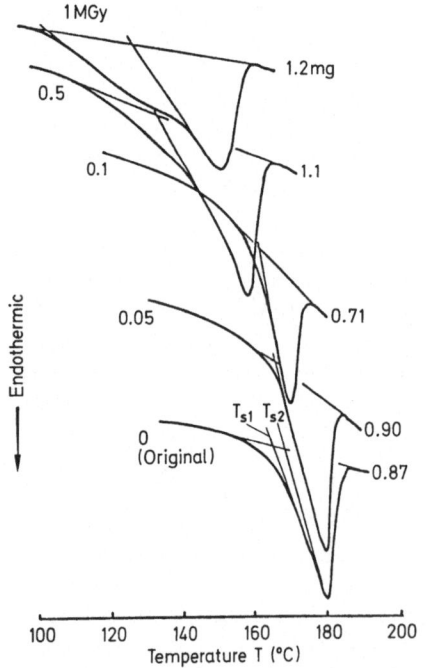

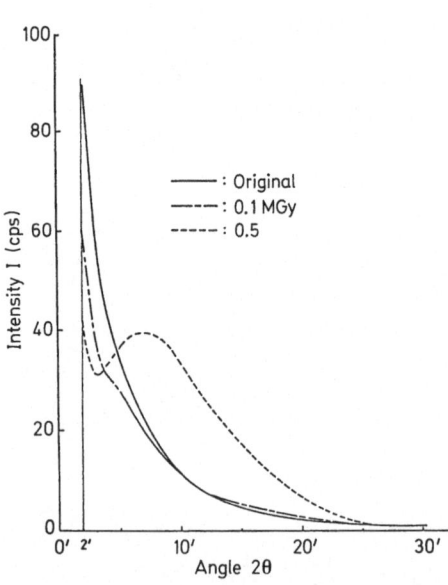

Fig. 5.11. Heating curves of poly(tetroxocane) (obtained at 84 °C) irradiated with various doses: Heating rate, 16 °C/min. Doses and sample weight are indicated in the figure

Fig. 5.12. Small-angle X-ray scattering curves of poly(tetroxocane) obtained at 84 °C in the meridional direction

in the irradiated sample as shown in the case of the sample prepared at 105 °C. On the other hand, SAXS behavior of PT_EOX prepared at 84 °C in the meridian is slightly different from that of PT_EOX prepared at 105 °C.

Figure 5.12 shows the SAXS intensity curves of PT_EOX prepared at 84 °C (similar to Fig. 5.11). The intensity curve of the as-polymerized sample decreases monotonically except of a small shoulder at ca. 5'. This small shoulder becomes clearer by irradiation with 0.1 MGy and forms a peak with 0.5 MGy. The peak position of ca. 7' corresponds to a long spacing of ca. 800 Å, indicating the periodical appearance of voids along the c-axis.

Figure 5.13 shows the electron microphotograph of as-polymerized PT_EOX irradiated with 1.0 MGy. The existence of a rippled feature perpendicular to the fibre axis is observed with a period of ca. 600 Å, indicating the periodical formation of voids as mentioned in the case of irradiated PTOX.

It is assumed that the polymer crystal of PT_EOX has a region accessible to radiation, i.e., the region where voids are easily formed by radiolysis. Such a region must be stressed by misfitting of polymer chains in the crystal lattice during the polymerization reaction, as mentioned in the case of PTOX.

Next, the effect of the irradiation on the WAXS profiles of polymers is examined.

Figure 5.14 shows the 009 reflection profiles of PTOX, PT_EOX, and drawn Delrin. The profile of PTOX is very sharp, the width being related to a crystal size of about 500 Å along the chain axis. The broad width in the profile of Delrin is caused by the large thickness (about 200 Å) of the lamellae. On the other hand, a extremely broad width, and an asymmetric shape, in the profile of PT_EOX, suggest a superposition due to the two types of polymer crystals.

The peak position of the profiles of PTOX and PT_EOX is quite similar to situate at a higher angle of 2θ than that of Delrin.

Figure 5.15 shows the WAXS profiles of as-polymerized PT_EOX, annealed at 170 °C, and irradiated. The asymmetrical feature of the as-polymerized sample is

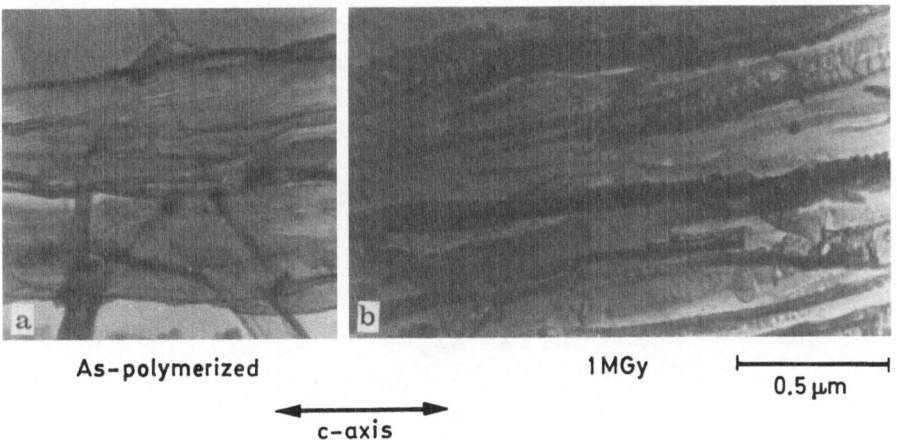

As-polymerized   1 MGy |———————|
                                                      0.5 μm

c-axis

Fig. 5.13a and b. Electron micrographs of as-polymerized (0 MGy) poly(tetroxocane) (obtained at 84 °C) irradiated with 1.0 MGy

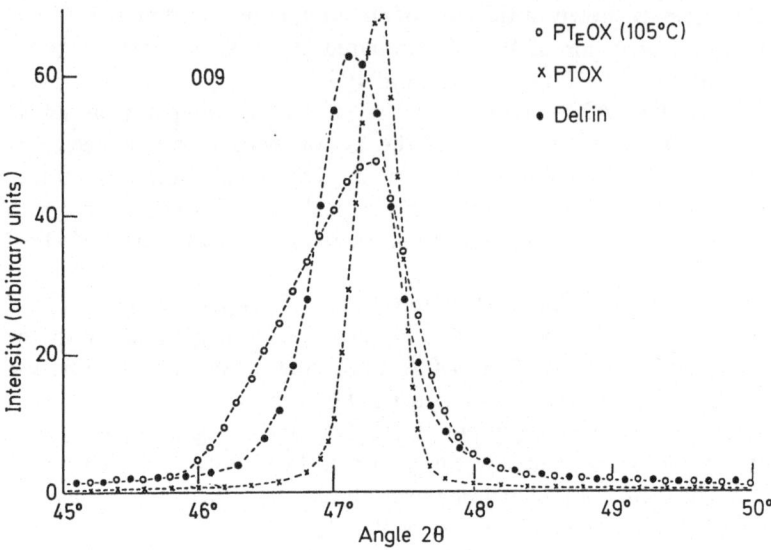

Fig. 5.14. Wide-angle X-ray scattering profiles of 009 reflections of poly(tetroxocane), poly(trioxane), and Delrin

changed into a symmetrical one by both treatments of annealing and irradiation, though the peak position of the annealed sample shifts to a lower-angle side, while that of the irradiated one remains constant. These facts suggest that the annealing causes a rearrangement of polymer chains to form thicker lamellae, while the irradiation causes radiolysis at the non-crystalline region. The change from an

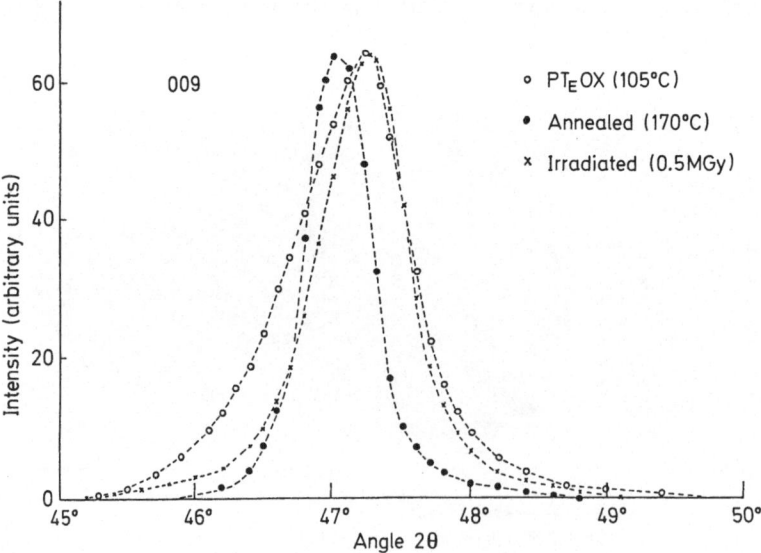

Fig. 5.15. Wide-angle X-ray scattering profiles of 009 reflections of poly(tetroxocane), as-poly-merlzed (105 °C), annealed (170 °C), and irradiated (0.5 MGy)

asymmetrical to a symmetrical profile in the as-polymerized sample also suggests that radiolysis affects the lattice constant of a crystallite as a result of the reflection at a lower angle, but not at a higher angle.

Figure 5.16 shows the WAXS profiles of PT_EOX prepared at 80 °C, which shows no lamellar morphology as mentioned above. The profile of the 009 reflection (Fig. 5.16a) of this polymer exhibits an asymmetrical tail at the lower angle side like the polymer prepared at 105 °C (Fig. 5.15), but not clearly distinguished in the 0018 reflection (Fig. 5.16b). Irradiation as well as annealing change the asymmetry of the profile into a symmetrical form reducing the tail at the lower angle side; this behavior is quite similar to that of the polymer prepared at 105 °C.

Accordingly, the same irradiation effect is noticed in PT_EOX irrespective to the presence of the sub-crystal or the lamellae.

The narrowing of the half-width in WAXS profiles, or the increase of the lattice spacing by irradiation or annealing, may be attributed to a relaxation from internal strains as a result of chain scission or rearrangement of the molecular chains, respectively.

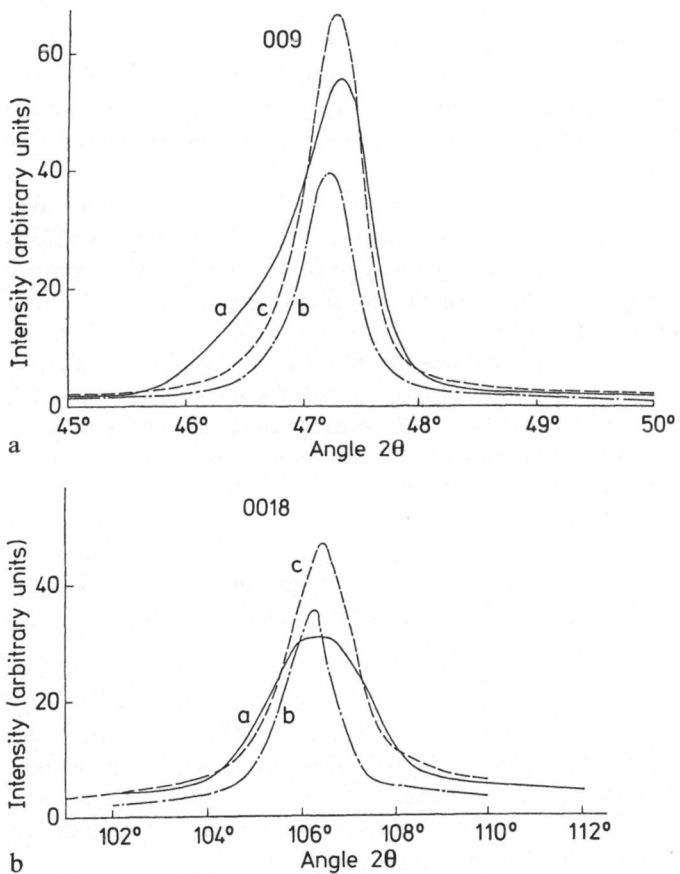

Fig. 5.16a and b. Wide-angle X-ray scattering profiles of 009 reflection (**a**) and 0018 reflection (**b**) of as-polymerized poly(tetroxocane) (80 °C) (a), annealed (160 °C) (b), and irradiated (0.5 MGy) (c)

Radiolysis or etching by irradiation may take place at a region which is strained or entropy restricted, quite similar as in the case of poly(trioxane) mentioned above.

6 Conclusion

The investigation of the X-ray scattering profiles of poly(tetroxocane) prepared by radiation-induced polymerization in the solid state suggest that two types of crystallites coexist in the polymers obtained above 80 °C. One is a lamellar type crystallite with a longitudinal dimension of 100 Å and fibrillar crystals with a crystallite size of 250 Å. Only fibrillar crystallites, as well as main and sub-crystals, were found in polymers obtained below 80 °C. The lamellar crystallites were less distorted than the fibrillar ones. In the case of poly(trioxane), main and sub-crystallites are observed; small amounts of randomly oriented crystals are also noticed.

The investigation of the melting behaviors of poly(tetroxocane) suggest the co-existence of at least two types of crystallites since a double endotherm is observed in the heating curves. On the other hand, poly(trioxane) gives a double endotherm in the heating curves, but an endotherm at higher temperature is due to superheating caused by entropy restriction.

The rearrangement of the polymer chains of the as-polymerized sample takes place by annealing at temperatures below the melting point. The annealing effects on the long spacing of 100 Å and the melting behavior of poly(tetroxocane) confirm the coexistence of lamellar crystallites which are situated between the fibrillar bundles of the extended chain crystallite.

Radiolysis reveals the characteristic features of polymers, where chain scission by irradiation occurs at a selective point. Poly(trioxane) forms a layer-like void, suggesting periodical characteristics of about 1000 Å along the c-axis of the polymer crystals. In the case of poly(tetroxocane) similar results are observed. The mean crystallite lengths of fibrillar crystals are obtained as about 500 Å in poly(trioxane) and 300 Å in poly(tetroxocane).

The fibrillar crystallites in polymers contain a region which is accessible to radiation, indicating the existence of polymer chain aggregation with stress or entropy restriction accumulated somewhat periodically during the polymerization in the solid state.

Acknowledgement: The authors wish to express their thanks to Mr. Teruo Ishibashi of Hokkaido University, and Dr. Osamu Yoda of JAERI, Takasaki, for the helpful assistance in the X-ray diffraction measurements.

The authors are also indebted to the lare Professor Mitchel Shen, University of California, for encouragement to prepare this review.

The authors gratefully acknowledge the critical reading of this manuscript by Professor J. B. Lando, Case Western Reserve University.

7 References

1. Okamura, S., Hayashi, K., Nakamura, Y.: Isotop. Radiat. *3*, 416 (1960)
2. Hayashi, K., Kitanishi, Y., Nishii, M., Okamura, S.: Makromol. Chem., *47*, 237 (1961)
3. Okamura, S., Hayashi, K., Kitanishi, Y.: J. Polym. Sci., *58*, 925 (1962)
4. Hayashi, K., Ochi, H., Nishii, M., Miyake, Y., Okamura, S.: J. Polym. Sci., *B1*, 427 (1963)
5. Nishii, M., Hayashi, K., Okamura, S.: J. Polym. Sci., *B7*, 891 (1969)
6. Hirschfeld, F. L., Schmidt, G. M.: J. Polym. Sci., A-1, *2*, 2181 (1964)
7. Chatani, Y., Kitahama, K., Tadokoro, H., Yamauchi, T., Miyake, Y.: J. Macromol. Sci.-Phys., *B4*, 61 (1970)
8. Chatani, Y., Uchida, T., Tadokoro, H., Hayashi, K., Nishii, M., Okamura, S.: J. Macromol. Sci.-Phys., *B2*, 567 (1968)
9. Amano, T., Fischer, E. W., Hinrichen, G.: J. Macromol. Sci.-Phys., *B3*, 209 (1969)
10. Wegner, G., Fischer, E. W., Munoz-Escalona, A.: Makromol. Chem., Suppl. *1*, 521 (1975)
11. a) Calson, J. P., Reneker, D. H.: J. Appl. Phys., *41*, 4296 (1970)
    b) Reneker, D. H., Calson, J. P.: ibid. *42*, 4606 (1971)
    c) Calson, J. P., Reneker, D. H.: ibid. *44*, 4293 (1973)
12. a) Nakase, Y., Kuriyama, I.: Polym. J., *4*, 517 (1973)
    b) Nakase, Y., Kuriyama, I., Odajima, A.: ibid. *8*, 35 (1976)
    c) Nakase, Y., Kuriyama, I., Nishijima, H., Odajima, A.: Japan. J. Appl. Phys., *16*, 1417 (1977)
13. a) Kato, T., Nakase, Y., Yoda, O., Kuriyama, I., Odajima, A.: Polym. J., *8*, 331 (1976)
    b) Nakase, Y., Kato, T., Yoda, O., Kuriyama, I., Odajima, A.: Polym. J., *9*, 605 (1977)
    c) Nakase, Y., Hayakawa, N., Kuriyama, I., Odajima, A.: Japan. J. Appl. Phys., *16*, 1703 (1977)
    d) Odajima, A., Ishibashi, T., Nakase, Y., Kuriyama, I.: Polym. J., *15*, 331 (1983)
14. a) Nakase, Y., Kuriyama, I., Nishijima, H., Odajima, A.: J. Mater. Sci., *12*, 1443 (1977)
    b) Odajima, A., Ishibashi, T., Nakase, Y., Kuriyama, L.: ibid. *13*, 77 (1978)
15. Iguchi, M.: Br. Polym. J., *5*, 195 (1973)
16. Odajima, A., Yamane, S., Ishibashi, T.: Rept. Prog. Polym. Phys. Japan, *21*, 147 (1978)
17. Hayashi, K., Nishii, M., Okamura, S.: J. Polym. Sci., *C4*, 839 (1963)
18. Statton, W. O.: J. Polym. Sci., *58*, 205 (1962)
19. Hosemann, R., Wilke, W.: Makromol. Chem., *118*, 230 (1968)
20. Nakase, Y., Yoshida, M., Ito, A., Hayashi, K.: J. Polym. Sci., A-1 *9*, 465 (1971)
21. Iguchi, T.: Makromol. Chem., *177*, 549 (1976)
22. Sakamoto, M., Ishigaki, I., Kumakura, M., Yamashina, H., Iwai, I., Ito, A., Hayashi, K.: J. Macromol. Chem., *1*, 639 (1966)
23. Nakase, Y., Yoshida, M., Ito, A., Hayashi, K.: J. Polym. Sci., A-1 *10*, 2181 (1972)
24. Harland, W. G., Khadr, M. M., Peters, R. H.: Polymers, *13*, 13 (1972)
25. a) Blais, J. J. B. P., Manley, R. st. J.: J. Macromol. Sci.-Phys., *B1*, 525 (1967)
    b) Holdworth, P. J., Fischer, E. W.: Makromol. Chem., *175*, 2637 (1974)
26. Hoffman, J. D., Weeks, J. J.: J. Res. NBS, *66A*, 13 (1962)
27. For example: Jaffe, M., Wunderlich, B.: Kolloid-Z. Z. Polym., *216/217*, 203 (1967)
28. Torikai, S.: J. Polym. Sci., *A2*, 239 (1969)
29. Keller, A., Willmouth, F. M.: J. Macromol. Sci.-Phys., *B6*, 493 (1972)
30. Miyagi, A., Wunderlich, B.: J. Polym. Sci., A-2 *10*, 1401, 2073, and 2085 (1972)
31. Zachmann, H. G.: Kolloid-Z. *216/217*, 180 (1967)
32. Kusano, T., Sutoh, Y., Murakami, K.: Nippon Kagaku Zasshi, *1974*, 1128 (1974)
33. Odajima, A., Ishibashi, T., Miyamoto, M.: Rept. Prog. Polym. Phys. Japan, *17*, 205 (1974)
34. Carazzolo, G., Lighissa, J., Mammi, M.: Makromol. Chem., *60*, 171 (1963)
35. a) Tsvankin, D. Ya.: Vysokomol. Soedin., *6*, 2078 and 2083 (1964)
    b) Tsvankin, D. Ya., Zubov, Yu. A., Kitaigorodskii, A. I.: J. Polym. Sci., *C16*, 4081 (1968)
36. Blundell, D. J.: Acta Cryst., *A26*, 472 (1970)
37. Crist, B.: J. Polym. Sci., A-2 *11*, 635 (1973)
38. Buchanan, D. R.: J. Polym. Sci. A-2 *9*, 645 (1971)

S. Okamura (Editor)
Received December 30, 1983

Molecular Interpretation of the Moduli
of Elastomeric Polymer Networks of Known Structure

J. P. Queslel and J. E. Mark
Department of Chemistry and the Polymer Research Center,
The University of Cincinnati, Cincinnati, Ohio 45221, USA

The most general molecular theory of rubberlike elasticity is reviewed as a preliminary to its use in the interpretation and understanding of the moduli of model elastomeric materials, i.e., polymer networks prepared in such a way that their structure is relatively well known. Applications are made to both perfect and imperfect networks, with critical evaluation of evidence possibly attributable to equilibrium elastic contributions from interchain entanglements. The relevance and importance of branching theory is covered in some detail.

Advances in Polymer Science 65
© Springer-Verlag Berlin Heidelberg 1984

1 Introduction

The understanding of the elastic properties exhibited by elastomer networks has been under investigation for a long time. The first molecular theories of rubber elasticity dealt with either an idealized phantom network which chains may move freely through one another [1,2], or a network in which the junction points move linearly ("affinely") with the macroscopic dimensions of the sample [2]. A refinement has been the introduction by Flory of the effects of constraints on crosslinks due to the neighboring chains [3,4], with such constraints depending on the network structure and deformation. A great improvement in the experimental testing of these theories is the use of model networks [5-9]. The specific chemical reactions used in their preparation leads to some quantitative information such as the molecular weight between crosslinking points and the functionality of the crosslinks. Nevertheless some controversies remain in the interpretation of experimental data, mainly due to concern about the completeness of the crosslinking reaction; in particular it has been argued that a contribution to the stress could arise from trapped inter-chain entanglements [10]. In this paper, after a review of the molecular theories, some experimental data are analyzed in term of the structure factors which relate the network modulus to degree of crosslinking and then are used to obtain values for the parameters of the most general Flory theory of rubberlike elasticity. Some arguments suggesting a contribution arising from trapped entanglements are discussed but the results are now treated with a modification of the usual Langley plot. Also the confusion between effective and active chains and junctions is clarified.

2 Review of Molecular Theory

2.1 The Phantom Limit

This network is composed of Gaussian chains which introduce forces between pairs of junctions so connected. The junctions fluctuate around their mean positions due to their Brownian motion. The instantaneous distribution of chain vectors r is not affine in the strain because it is the convolution of the distribution of the mean vector $\bar{r}$ (which is affine) with the distribution of the fluctuations Δr (which are independent of the strain).

Flory [11] has rederived the elastic free energy for such a network using its cycle rank ξ, which is the number of chains which have to be cut to reduce the network to an acyclic structure or tree [12]. Specifically,

$$\Delta A_{el}(ph) = (1/2)\,\xi kT(I_1 - 3) \tag{1}$$

where I_1 is the first invariant of the tensor of deformation, k the Boltzmann constant, and T the absolute temperature.

Networks consist of ν linear chains whose ends are joined to multifunctional junctions of any functionality $\varphi > 2$ or to junctions of any combination of functionalities $\varphi \geq 2$, at least some of which have functionalities $\varphi > 2$. (In this paper, the functionality will be designated by either φ or f, to follow the two widespread notations). The cycle rank is the difference between the number ν of effective chains and the total number μ of effective junctions of any functionality $\varphi \geq 2$:

$$\xi = \nu - \mu + 1 \simeq \nu - \mu \tag{2}$$

Scanlan [14] and Case [15] have defined an active junction as one joined by at least three paths to the gel network and an active chain as one terminated by an active junction at both its ends. Graessley [16] and Flory [13] have proved that

$$\nu - \mu = \nu_a - \mu_a \tag{3}$$

where ν_a and μ_a are the number of active chains and active junctions, respectively. Equation (3) is the link between Flory's treatment and another due to Graessley [16, 17]. The shear modulus can now be shown to be

$$G_{ph} = (\nu - \mu)\,RT = (\nu_a - \mu_a)\,RT \tag{4}$$

where R is the gas constant.

For a perfect network of functionality φ [17]

$$\nu_a - \mu_a = \left(1 - \frac{2}{\varphi}\right) \nu_a \tag{5}$$

or

$$\frac{\mu_a}{\nu_a} = \frac{2}{\varphi}$$

2.2 The Affine Limit

2.2.1 General Comments

For a tetrafunctional phantom network, the mean square fluctuations of r are half of $\langle r^2 \rangle_0$, the mean square of the end-to-end separation of the free uperturbed chains [11].

In a real network, diffusion of the junctions about their mean positions may be severely restricted by neighboring chains sharing the same region of space. The extreme case is the affine network where the fluctuations are completely suppressed; in this case the instantaneous distribution of chain vectors is affine in the strain.

The elastic free energy has the expression

$$\Delta A_{el}(\text{aff}) = (1/2)\,\nu kT(I_1 - 3) - (\nu - \xi)\,kT \ln (V/V_0) \tag{6}$$

where V and V_0 are the actual and reference volumes, respectively. In a perfect network of uniform functionality $\varphi > 2$

$$\xi = \nu_a - \mu_a = \left(1 - \frac{2}{\varphi}\right) \nu_a \tag{7}$$

Thus if the second term of Eq. (6) is temporarily omitted

$$\Delta A_{el}(aff) = (1/2) \, \nu_a kT(I_1 - 3) \tag{8}$$

and the shear modulus is then simply

$$G_{aff} = \nu_a RT \tag{9}$$

Or in a real, imperfect network

$$G_{aff} = \nu RT \tag{10}$$

It is worth noting that $G_{aff} > G_{ph}$, which is a consequence of the fact that the fluctuations of junctions in a phantom network are unaffected by deformation. Flory [13] has recently pointed out that the identification of ν_a with the number of effective chains ν is valid only for perfect networks (see Sect. 6.1). For an imperfect tetrafunctional network, Flory [18] has shown that

$$\nu = \nu_0 \left(1 - 2\frac{M_c}{M_n}\right) \tag{11}$$

where ν_0 is the total number of chains in the network, M_c is the average molecular weight of a network chain and M_n is the average number molecular weight of the primary molecules.

It is necessary to emphasize that the common use of ν_a instead of ν in Equation (10) for end-linked model networks [19] is only an approximation when the networks are formed with nonstoichiometric ratio between the crosslinking molecules and end-reactive chains. Nevertheless this approximation seems not to be serious [14] and we shall see that it is certainly valid for high functionality networks.

2.2.2 Swelling Equilibrium

The maximum degree of swelling of a network characterized by the volume fraction v_{2m} of polymer and the number density ν of chains are related in the Flory theory [20−22] by the Equation

$$\nu = - \left[\ln (1 - v_{2m}) + v_{2m} + \chi_1 v_{2m}^2\right]/V_1 v_{2s}^{2/3} \left(v_{2m}^{1/3} - \frac{2}{\varphi} v_{2m}\right) \tag{12}$$

where V_1 is the molar volume of the solvent, v_{2s} is the volume fraction of polymer in the system during the crosslinking procedure and χ_1 is the interaction parameter between polymer and solvent [20, 23].

2.3 Real Networks

2.3.1 Stress-Strain Measurements in Elongation

Experimental measurements of stress versus the macroscopic deformation λ in uniaxial extension have revealed that the behavior of real networks is between the affine and phantom limits.

Moreover, the representation of the reduced force (modulus) versus λ^{-1} is well-fitted by a straight line in the region $\lambda^{-1} = 0.9 - 0.5$. The most used definition of the modulus in elongation is

$$[f^*] \equiv f^* v_2^{1/3}/(\lambda - \lambda^{-2}) \tag{13}$$

where f^* is the nominal stress f/A^0, f is the equilibrium value of the elastic force, A^0 is the undeformed cross-sectional area of the sample, and v_2 is the volume fraction of polymer in the network (which is frequently swollen).

If the slope of this straight line is designated by $2C_2$ and its extrapolation at infinite deformation $(\lambda^{-1} = 0)$ by $2C_1$, then one obtains the phenomenological Mooney Eq. [24]

$$[f^*] = 2C_1 + \frac{2C_2}{\lambda} \tag{14}$$

The constants $2C_1$ and $2C_2$ have already been obtained in a great number of different systems [25], but no convincing molecular explanation had been found until recently. The first theoretical attempts to account for this dependence of $[f^*]$ on λ in terms of a gradual transition between the two extreme types of network deformation are due to Ronca and Allegra [26] and Flory [11]. A great achievement has been a recent molecular theory due to Flory [3, 11, 27-29]. In this model, the restrictions on the fluctuations of junctions due to the neighboring chains are repressented by domains of constraints. At small deformations, the stress is enhanced relatively to that exhibited by a phantom network. At large strains, or at high dilation, the effects of restrictions on fluctuations vasnish and the relationship of stress to strain converges to that for a phantom network. The topological and mathematical treatment leads to the expression

$$[f^*] = f_{ph}\left[1 + \frac{f_c}{f_{ph}}\right] \tag{15}$$

where f_{ph} is the usual phantom modulus, and f_c/f_{ph} is the ratio of the force due to entanglement contraints to that for the phantom network. In the latest theoretical refinement, this ratio depends on two parameters, the most important being $\varkappa$, which measures the severity of the entanglement constrains relative to those imposed by the phantom network. Another parameter ζ takes into account the non-affine transformation of the domains of constraints with strain. The specific expression for f_c/f_{ph} in uniaxial extension is

$$f_c/f_{ph} = \frac{\mu}{\xi}[\lambda K(\lambda_1^2) - \lambda^{-2}K(\lambda_2^2)]/(\lambda - \lambda^{-2}) \tag{16}$$

where $\lambda_1 = \lambda$, $\lambda_2 = \lambda^{-1/2}$ and μ/ξ is the ratio of the number of effective junctions to the cycle rank. [For a perfect network $\mu/\varsigma = 2/(\varphi - 2)$]. The function $K(x^2)$ is given by

$$K(x^2) = B[\dot{B}(B + 1)^{-1} + g(\dot{g}B + g\dot{B})(gB + 1)^{-1}] \tag{17}$$

with

$$g = x^2[1/\varkappa + \zeta(x - 1)]$$

$$\dot{g} = 1/\varkappa - \zeta(1 - 3x/2)$$

$$B = (x - 1)(1 + x - \zeta x^2)/(1 + g)^2$$

$$\dot{B} = B[(2x(x - 1))^{-1} - 2\dot{g}(1 + g)^{-1} + (1 - 2\zeta x)(2x(1 + x - \zeta x^2))^{-1}]$$

The typical plot of $[f^*]$ versus λ^{-1} is a sigmoidal curve (see Fig. 5–7 discussed below). Experimental data in elongation and compression fitted by this theory and also discussed below indicate values of $\varkappa$ of the order 5–10 [30-34].

2.3.2 Swelling Equilibrium

Flory [34] has derived the elastic free energy of dilation of a network with account of restrictions of fluctuations of junctions. Quantitative agreement has been reported for vapor sorption measurements. Particularly impressive is reproduction of the observation that the product of the linear expansion ratio λ and the elastic contribution $(\mu_1 - \mu_1^0)_{el}$ to the chemical potential of the diluent in a swollen network exhibits a maximum with increase in λ, which is contrary to previous theory [35].

It is convenient to compare the phantom modulus obtained by stress-strain measurements to that obtained from swelling equilibrium studies [32, 36]

$$f_{ph} = - (RT/V_1)[\ln(1 - v_{2m}) + v_{2m} + \chi_1 v_{2m}^2]v_{2m}^{-1/3}/[1 \\ + (\varphi/2 - 1)^{-1} K(v_{2m}^{-2/3})] \tag{18}$$

where K is defined by Eq. (17). The other parameters are the same as in Eq. (12). One has to recognize that the parameter $\varkappa$ used in the formula of the function K is perhaps not exactly the same as the one determined by stress-strain measurements of unswollen samples, since the effect of swelling may not be only to deform isotropically the sample, but also could increase the fluctuations of junctions by a dilution effect.

Some other models have been proposed to account for the behavior of real networks [37], but they don't seem to reproduce as completely the experimental observations [38].

3 The Use of Model Networks

Elastomers are generally crosslinked in a random manner and therefore it is difficult to obtain the quantitative, independent information (molecular weight M_c between crosslinks, functionality φ of the network) required to test the molecular theories. However, some systems such as natural rubber cured by dicumyl peroxide [39-42], irradiated polymers [43-46] or even chemically randomly crosslinked polydimethylsiloxane [38,47] and poly(ethyl acrylate) [32] have been investigated. Mark has exploited a powerful tool in the study of rubber elasticity [6], specifically the use of model networks generally formed by the crosslinking reaction of functionally-terminated chains of known molecular weight with end-linking molecules of known functionality. An example of such a technique is the end-linking of hydroxyl-terminated polydimethylsiloxane with alkoxy silane or triisocyanate [48]. Model polyurethane [49, 50] or polyisobutylene [51] networks have also been prepared in this way. These model should have values of M_c essentially equal to the number-average molecular weight M_n of the chains prior to their end-linking and values of φ equal to the functionality of the end-linking agent. Furthermore, the wide variety of available reactants enables the study of networks of very different M_c [52-54], different functionalities [55, 56], dangling-chain irregularities [57], comb-like crosslinks [58], bimodal networks [59], networks of low extents of crosslinking [60] and networks prepared at different degrees of dilution [61]. While most of the experiments have been performed in uniaxial extension (or compression) at equilibrium, others were carried out in uniaxial extension at different speeds [62], or by viscoelasticity measurements [63, 64]. The development of the modulus and the extent of the crosslinking reaction have also been followed continuously [65].

4 Experimental Data

4.1 Completeness of the Cross-Linking Reaction

The model networks are generally prepared with end-reactive oligomers B_2 and crosslinking molecules A_f of functionality f [the $(A_f + B_2)$ system]. A perfect network is obtained for a theoretical value of unity for the composition ratio

$$r = \frac{f}{2}\left(\frac{[A_f]_0}{[B_2]_0}\right) \tag{19}$$

where $[A_f]_0$ and $[B_2]_0$ are the initial molar concentrations of crosslinking agents and oligomers, respectively. Macosko has argued that it is better to use a value of r a little larger than unity ($\simeq$ 1.1–1.2) since a maximum of the shear modulus is obtained for these values [54], but this should only lead to more imperfect networks of unknown topology [66]. A possible determination of the extent of reaction is the iodometric titration of unreactive species in the case of vinyl-crosslinked PDMS [67], but the quantity of unreacted vinyl groups is too small to allow accurate measurements. Infra-red analysis was also performed [68] but it is difficult to use this method to obtain quantitative results. Another commonly used method is the measurement of the sol fraction, but this sol fraction contains not only unreactive species, but also low molecular weight oligomers, cyclics, and catalyst. For this reason, Macosko and Merrill have extensively used gel permeation chromatography to analyze the initial products and the sol fraction [54, 56]. A corrected value of the sol fraction is then calculated and from this value the extent of reaction and different structural parameters of the network are derived by branching theory. However the main assumption of branching theory is that all of the species in the sol fraction are reactive and Mark [51] has recently pointed that the formation of cyclics [an occurrence supported by simulation of the end-linking reaction [66, 69)] could give a misleadingly high value of the sol fraction even in a network containing essentially no imperfections.

At this point of the discussion, it is necessary to emphasize that the quantities involved in the measurements of sol fraction are small and it is difficult to obtain accurate results. It is also worthy noting that Kennedy and coworkers have recently prepared perfect polyisobutylene networks with extremely low sol fractions [51, 70]. Some analyses of sol fractions by GPC [71] and determinations of the crosslinking density by ^{29}Si solid-state NMR [72] are now in progress in an attempt to obtain a better characterization of polydimethylsiloxane networks. A final observation is that it seems that the longer the primary chains are, the more difficult is the achievement of complete reaction, and the higher the sol fraction. This can be easily understood in terms of problems regarding the diffusion of reactive species in a gel [74].

4.2 Structure Factors Obtained in Elongation Studies

4.2.1 General Comments

We have already mentioned that a widely used plot of the experimental results in uniaxial extension is the Mooney one. The straight line obtained by fitting the experimental points with a least-squares analysis can be extrapolated to infinite deformation ($\lambda^{-1} = 0$) and the value of the reduced force is then $2C_1$. In a similar manner, the extrapolation to zero deformation ($\lambda^{-1} = 1$) leads to ($2C_1 + 2C_2$) which is not far from the shear modulus G recorded by measurements at small deformations. The constant $2C_2$ is the slope of the line. An extensive review of $2C_1$ and $2C_2$ by Mark [25] on a great number of elastomeric networks didn't yield an unambigous molecular explanation of these two parameters. Both Ferry [75, 76] and Graessley [45, 78], however, have found some correlations. As proposed by the latter author, $2C_2$ accounts for approximately half of the topological contribution and contains essentially no contribution from the chemical network, and $2C_1$ consists of the entire chemical phantom network contribution, plus the other half of the topological contribution. The scattering of the data, however makes these conclusions doubtful. As already mentioned, account of the increase in non-affineness [11, 26] with increase in deformation is very successful in explaining the experimental results.

4.2.2 High-Deformation Limit

For a perfect end-linked network of functionality φ, the phantom modulus can be obtained from Eq. (5):

$$f_{ph} = \left(1 - \frac{2}{\varphi}\right) v_a RT \qquad (20)$$

where R is the gas constant and T is the absolute temperature. The quantity v_a is number density of elastically active strands which, for a perfect network formed in the unswollen state, is also the number density of primary chains

$$v_a = \frac{\varrho}{M_n} \qquad (21)$$

where ϱ is the density of the polymer and M_n is the number-average molecular weight of the primary chains. Hence

$$f_{ph} = \left(1 - \frac{2}{\varphi}\right) \frac{\varrho RT}{M_n} \tag{22}$$

and the structure factor at infinite deformation is given by [48]

$$A_\varphi = \frac{2C_1}{\varrho RT/M_n} \tag{23}$$

It is therefore interesting to compare A_φ and $\left(1 - \frac{2}{\varphi}\right)$.

Figures 1 and 2 show the dependence of the structure factors A_3 and A_4, respectively, on the number-average molecular weight of primary chains, which is also the molecular weight between crosslinks M_c. The results shown are only those for networks prepared with a stoichiometric ratio $r = 1$ [see Eq. (19)] of the crosslinking molecules and the reactive oligomers. The crosslinking reaction is assumed to be complete. For trifunctional and tetrafunctional networks, $1 - 2/\varphi = 0.33$ and 0.50 respectively. In Fig. 1, A_3 has a value near 0.33 only for M_n less than $10^4\,\mathrm{g\,mol^{-1}}$, and for higher molecular weights $A_3 > 0.33$. In Fig. 2, A_4 is always higher than 0.50. The numerical values are reported in Tables 1 and 2. Some authors have argued that this discrepancy between A_Φ and $1 - 2/\varphi$ is due to topological contributions ("inter-chain entanglements") [56,75,77]. This conclusion is based on the assumption that $2C_1$ is the actual network modulus at infinite deformation. Flory [3] has pointed out that $2C_1$ is the extrapolation of a straight line which is only a small part (in a narrow range of deformations) of a more general sigmoidal curve. Due to the use of λ^{-1} as independent variable, the Mooney plot gives more importance to small deformations and generally this extrapolation

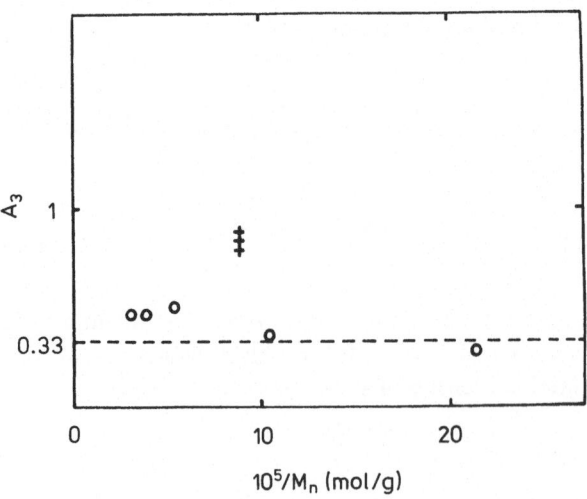

Fig. 1. Dependence of the structure factor at infinite deformation on M_n for perfect trifunctional end-linked PDMS networks. The experimental points are due to Mark [52] (O) and Llorente [55] (+)

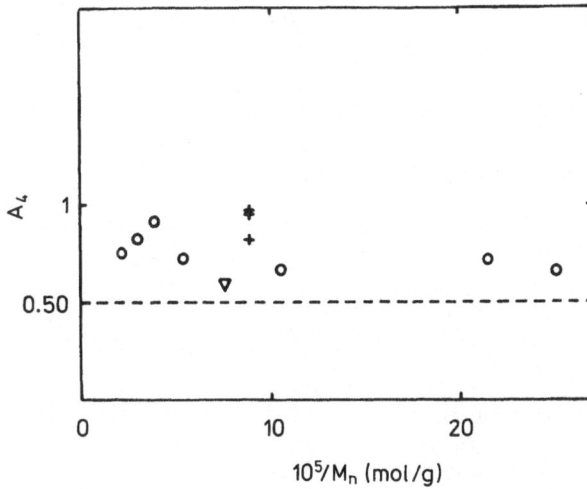

Fig. 2. Dependence of the structure factor at infinite deformation on M_n for perfect tetrafunctional end-linked PDMS networks. Mark [6] ($\bigcirc$); Llorente [54] ($+$); Falender [75] (∇)

Table 1. Structure factors for perfect trifunctional end-linked PDMS networks (T = 298 K)

Sources	M_c g mol^{-1}	A_3	A_3'	$10^5 M_c^{-1}$ (g mol^{-1})$^{-1}$
Mark [52]	4000	0.31	0.34	25.00
	4700	0.29	0.31	21.28
	9500	0.36	0.59	10.53
	18500	0.51	0.97	5.41
	25600	0.47	1.01	3.91
	32900	0.47	0.92	3.04
Valles [65]	11600		1.05^a	8.62
Meyers [68]	8800		0.76	11.36
	11100		0.77	9.01
	15200		1.28^b	6.58
	21600		1.51^c	4.63
Llorente [55]	11300	0.88	1.32	8.85
		0.85	1.32	
		0.79	1.37	
Macosko [54]	22400		0.57^d	4.46
	11400		0.59^d	8.77

^{a} r = 0.997, T = 300 K; ^{b} r = 1.01; ^{c} r = 0.996; ^{d} r = 0.998

doesn't give the origin of the sigmoidal curve. If we suppose now that a topological contribution is added to the junction contribution, the resulting curve will also have a complicated shape and the problem remains the same. A last argument could be the value of $2C_1$ a little less than 0.33 for perfect trifunctional polyisobutylene networks [51] which proves definitively the nonphysical meaning of $2C_1$ because it can't be less than the phantom modulus. (It is obvious that this statement is valid only on the assumption of perfect networks). For the same reasons, it is probably best to postpone detailed analysis of the influence of the functionality on A_φ (Table 3). An important final remark concerns elongation measurements on swollen samples. In particular,

Table 2. Structure factors for perfect tetrafunctional end-linked PDMS networks (T = 298 K)

Sources	M_c g mol^{-1}	A_4	A'_4	$10^5 M_c^{-1}$ (g mol^{-1})$^{-1}$
Falender [75]	13000	0.59	0.97	7.69
Mark [6]	4000	0.66	0.70	25.00
	4700	0.71	0.77	21.38
	9500	0.67	0.87	10.53
	18500	0.72	1.03	5.41
	25600	0.91	1.50	3.91
	32900	0.82	1.41	3.04
	45000	0.75	1.34	2.22
Valles [65]	11600		1.24[a]	8.62
			1.17[b]	
Llorente [55]	11300	0.94	1.34	8.85
		0.82	1.37	
		0.96	1.32	
Macosko [54]	11400		0.97	8.77
	4190		0.60	23.9

[a] T = 300 K; [b] T = 313 K

Table 3. Dependence of the structure factors on functionality for perfect end-linked PDMS networks

Source	M_c g mol^{-1}	φ	A_φ[a]	A'_φ[a]	$10^5 M_c^{-1}$ (g mol^{-1})$^{-1}$
Llorente [55]	11300	3	0.84	1.34	8.85
		4	0.91	1.34	
		4.6	1.05	1.42	
		5	1.08	1.34	
		6	1.15	1.41	
		8	1.31	1.43	
		11	1.40	1.46	
		37	1.12	1.22	

[a] Average values for several samples

Flory and coworkers [32] have shown that $2C_1$ obtained with swollen networks of poly(ethyl acrylate) are in good agreement with the phantom modulus calculated exclusively from the chemical constitution of the networks.

4.2.3 Low-Deformation Limit

In an affine network, the fluctuations of junctions are completely suppressed. According to the Flory theory, the behavior of a real network is between the affine and the phantom limits, closer to affine near the isotropic state, and tending to the phantom one with increasing deformation. It is interesting to known the nature of the isotropic starting state. This is done with the structure factor at zero deformation which is defined as the ratio between the actual modulus of the network at $\lambda = 1$ and the theoretical affine modulus [Eq. (10)]

$$A'_\varphi = \frac{G}{\nu RT} \tag{24}$$

where for a perfect end-linked network of functionality φ [equation (21)]

$$\nu = \nu_a = \frac{\varrho}{M_n} \tag{25}$$

and G is the shear modulus at small deformations. The sum $2C_1 + 2C_2$ is frequently used as an approximation for G. In the next section it is shown that the error thus committed is from 1 to 10 percent of the value of A'_φ. Unlike $2C_1$, $2C_1 + 2C_2$ is a more nearly valid extrapolation because the experimental measurements are generally done at rather small deformations. Therefore, A'_φ should have a more reliable physical meaning. Values of A'_3 and A'_4 are reported in Tables 1 and 2 for end-linked PDMS networks presumed to be perfect, and their dependence on M_n is shown in Fig. 3 and 4.

If there is no entanglement contribution to the equilibrium values of the stress, A'_φ must be unity or less. Hence, A'_φ is generally compared with unity. In Fig. 3, for $M_n < 10^4$ g mol^{-1}, A'_3 is rather small and effectively less than one. It seems to increase with the molecular weight of the chains. It was already predicted by Flory that the severity of the restrictions on the fluctuations of junctions (accounted for by the parameter $\varkappa$) increases with M_n [29]. For $M_n > 10^4$, the scattering of data around unity doesn't permit any firm conclusion. In Fig. 4, for $M_n < 10^4$ A'_4 is also less than unity and higher than A'_3 at constant M_n. But it seems now that for large values of M_n, A'_4 might be equal to or greater than unity. However, the scattering of data is serious and this conclusion could be doubtful. The fact that 10^4 g/mol is the same order of magnitude as the molecular weight between entanglements in PDMS determined as by Ferry [78] may be fortuitous. Similar results were reported by Llorente [55] (Table 3) for networks of $M_n = 11\,300$. When the functionality increases, A' also

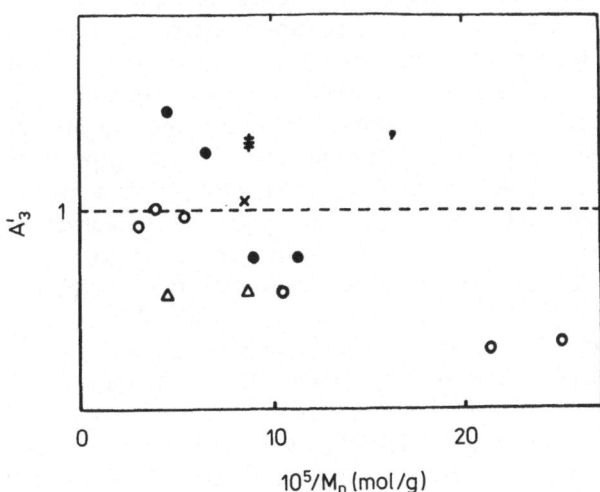

Fig. 3. Dependence of the structure factor at zero deformation on M_n for perfect trifunctional end-linked PDMS networks. Mark [52] (O); Llorente [55] (+); Macosko [54] ($\triangle$); Valles [65] ($\times$); Meyers [68] ($\bullet$)

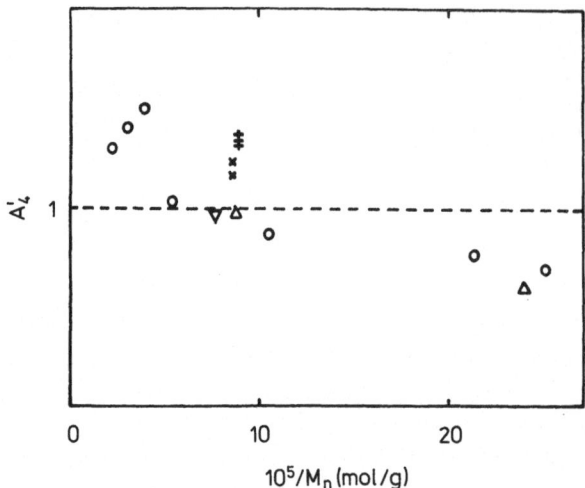

Fig. 4. Dependence of the structure factor at zero deformation on M_n for perfect tetrafunctional end-linked PDMS networks. Mark[6] ($\bigcirc$); Llorente[55] ($+$); Macosko[54] ($\triangle$); Valles[65] ($\times$); Falender[75] (∇)

increases. This is not surprising because fluctuations should be more restricted for junctions of high functionality. In the next section, we shall apply the Flory theory to determine the parameters $\varkappa$ and ζ for networks whose values of A'_φ are less than unity. In this case, it is not necessary to assume any entanglement contribution. Then the possible situation of $A'_\varphi > 1$ will be discussed. We can already notice that this occurs at large molecular weights and thus may be due only to non-equilibrium effects (difficulty in reaching elastic equilibrium).

5 Perfect End-Linked Networks

5.1 The Parameters $\varkappa$ and ζ

The Flory theory was already described in an earlier Section. Recently, Flory[29] has suggested that the parameter $\varkappa$ may be expected to vary inversely as the square root of the phantom modulus for perfect tetrafunctional networks (a result which may be generalized to other functionalities). This statement is based on the calculation of the degree of interpenetration of chains and junctions. The fitting of the elongational measurements to the Flory theory requires three parameters: the starting point of the curve at $\lambda^{-1} = 0$ (i.e. f_{ph}), $\varkappa$ and ζ. A reasonable assumption for perfect networks is to take f_{ph} equal to the theoretical value calculated from Eq. (22). The visualization of infinite deformation is not easy because it is always thwarted by the limited extensibility of chains. But the idea of fluctuations similar to phantom fluctuations (Flory) or of vanishing entanglements[42, 56, 60] supports the assumption of calculation of f_{ph} from the chemical constitution of the network. In a first approximation, ζ will be supposed to be zero; results discussed later seem to support this assumption. Thus, only $\varkappa$ has to be determined.

All the calculations presented in this paper (Flory theory, branching theory) were performed with a Digital Minc 23 VT 105 computer. For the fitting of the experimental

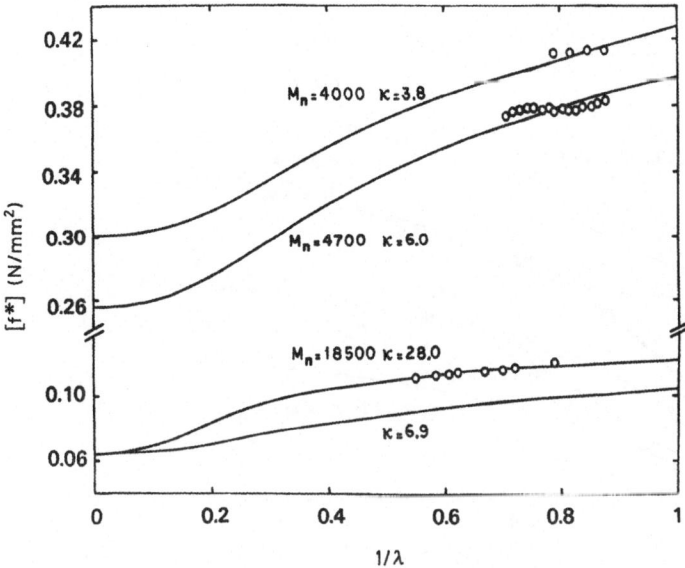

Fig. 5. Determination of the parameters of the Flory theory for perfect tetrafunctional networks: Mark [6] (○); M_n = 4000, 4700, 18500 g mol^{-1}

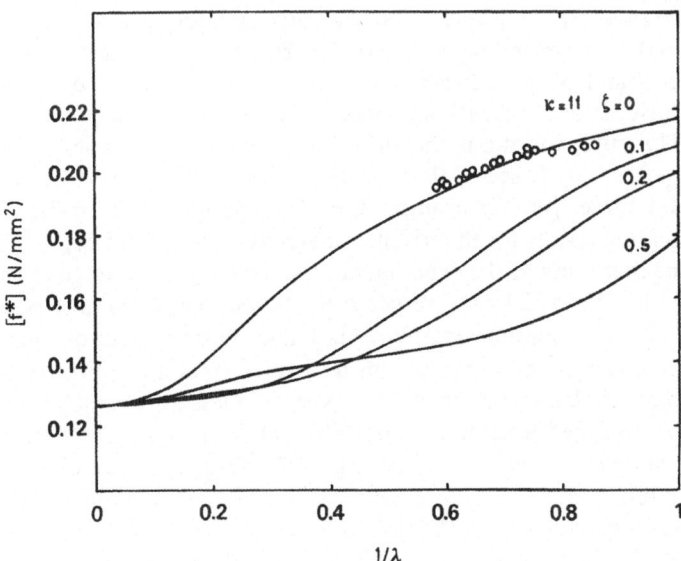

Fig. 6. Determination of the parameters of the Flory theory for perfect tetrafunctional networks: Mark [6] (○); M_n = 9500 g mol^{-1}. The results show the influence of ζ

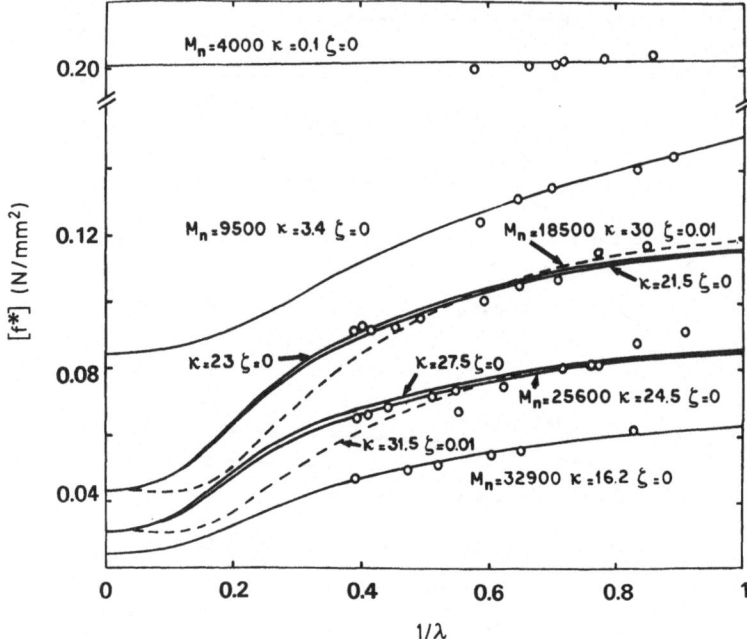

Fig. 7. Determination of the parameters of the Flory theory for perfect trifunctional networks: Mark [52] ($\bigcirc$); M_n = 4000, 9500, 18500, 25600, 32900 g mol^{-1}

points, the same error range was attributed to the reduced force at λ^{-1} = 0.6, 0.7, 0.8, 0.9. Then $\varkappa$ was chosen so as to minimize this error range. If then several values of $\varkappa$ were still possible, the minimization of the error range at λ^{-1} = 0.5 was used. The results are presented in Table 4 and the corresponding curves in Fig. 5–7. The fitting by this method is generally good, with the error range similar to the experimental one. In Fig. 5 and 6 the reduced forces are plotted against the inverse of deformation for tetrafunctional PDMS networks [6]. The experimental points can be fitted with ζ = 0. The influence of ζ is shown in Fig. 6; mainly it changes the curve in the region λ^{-1} = 0.5 — 1. It seems that the general slopes of the Mooney lines and the experimental curves [see for example Gee [79]] support $\zeta \simeq 0$. In Fig. 7 are plotted the corresponding results for the trifunctional networks [52]. For M_n = 18 500 and 25 600 g/mol, the scattering of the experimental points suggests a range of values of $\varkappa$, mainly around 21.5–23 and 24.5–27, respectively. In these cases, it is also possible to use a non-zero value for the parameter ζ. We obtain then a higher value of $\varkappa$ because the reduced force decreases when ζ departs from zero at constant $\varkappa$ (see Fig. 6), and $\varkappa$ has to increase to counterbalance this effect. However ζ is a less important parameter than $\varkappa$ and has to be assigned a value not too different from zero. The unrealistic value of $\varkappa$ for the sample having φ = 3 and M_n = 32900 g/mol may come from imperfections in this high molecular weight network. This is consistent with its value of A'_φ being less than those for M_n = 18 500 and 25 600 g/mol. In Tab. 4, it can be seen that the values of $\varkappa$ increase with the molecular weight of the chains and the functionality, which is in good agreement with Flory's predictions. However the product $\varkappa f_{ph}^{1/2}$ is not constant but increases with M_n and φ. A similar conclusion was

Table 4. Values of the Flory parameters $\varkappa$ and ζ for perfect trifunctional and tetrafunctional end-linked PDMS networks

Sources	φ	M_n g mol^{-1}	f_{aff} N mm^{-2}	f_{ph} N mm^{-2}	$1 - \dfrac{2}{\varphi}$	A'_φ	$\varkappa$	ζ	$kf_{ph}^{1/2}$	$A'_{\varphi,x}$	h	h_e
Mark [52]	3	4000	0.6019	0.2006	0.33	0.34	0.1	0	0.04	0.34	0.99	0.99
		4700	0.5122	0.1707		0.31[a]	~0				1	
		9500	0.2534	0.0845		0.59	3.4	0	0.99	0.59	0.61	0.61
		18500	0.1301	0.0434		0.97	23	0	4.79	0.90	0.15	0.15
							21.5		4.48	0.89	0.16	0.16
							30	0.01	6.25	0.91	0.13[c]	0.13
		25600	0.0940	0.0313		1.01[b]	27.5	0	4.87	0.91	0.13	0.13
							24.5	0	4.33	0.90	0.15	0.14
							31.5	0.01	5.57	0.93	0.13[c]	0.11
		32900	0.0732	0.0244		0.92	16.2	0	2.53	0.86	0.21	0.21
Mark [6]	4	4000	0.6019	0.3009	0.50	0.70	3.8	0	2.07	0.71	0.58	0.58
		4700	0.5122	0.2561		0.77	6.0	0	3.02	0.78	0.45	0.44
		9500	0.2534	0.1267		0.87	11.0	0	3.92	0.86	0.29	0.28
		18500	0.1301	0.0651		1.03[b]	28.0	0	7.14	0.95	0.13	0.12

[a] Since $A'_\varphi < 0.33$, the network was assumed to be phantom; [b] see remark in text; [c] by equation reported in the appendices

reached by Erman [29]. Now we can calculate the true structure factor at zero defor-mation $A'_{\varphi, \varkappa}$ which is the ratio of the Flory reduced force at $\lambda^{-1} = 1$ and the theoretical affine modulus f_{aff}. The resulting values of $A'_{\varphi, \varkappa}$ are listed in Table 4. The main departures from A'_{φ} occur at large M_n; this is probably due to the generally higher slopes of the Mooney lines for these samples and thus the extrapolation overestimates G. Thus the fitting of experimental curves with A'_{φ} slightly larger than unity becomes possible [samples designated by (b) in the Table]. In any case, the small difference (less than 8%) between A'_{φ} and $A'_{\varphi, \varkappa}$ permits the use of $(2C_1 + 2C_2)$ to calculate the structure factor at zero deformation. As already mentioned, Erman has also used the Flory theory, but his methods of fitting are quite different. He chose f_{ph} so as to make the experimental and theoretical reduced forces equal at $\lambda^{-1} = 1$, or used an arbitrary choice of f_{ph} and $\varkappa$, or in agreement with Pak [31] chose $\varkappa$ inversely proportional to $f_{\text{ph}}^{1/2}$. This is somewhat arbitrary and enables fitting data even when $A'_{\varphi} > 1$. His values of $\varkappa$ are generally smaller than the values reported here. Figure 5 shows the curve calculated with Erman's value $\varkappa = 6.9$ for the same sample ($M_n = 18\,500$) and with our assumption for f_{ph}.

It should be mentioned that this elegant new theory developed by Flory and Erman has the tremendous advantage of reproducing experimental measurements over the full range of deformations from large compressions to high elongations, as was shown by Pak [31].

5.2 The Parameter h

To account for the intermediate behavior between the affine and phantom modulus of a real network, Graessley [45] has proposed an expression for the small strain modulus G as a function of an empirical parameter having a value between 0 (affine) and 1 (phantom). Specifically,

$$G = (v_a - h\mu_a) RT \qquad (26)$$

where for a perfect network $v_a = v$ and $\mu_a = \mu$. Since h is widely used, it is interesting to derive it for the present results. If we call $G_\varkappa$ the extrapolation of the Flory curve at $\lambda^{-1} = 1$ and equate it to G, it can be shown by combination of Eq. (5), (25) and (26) that

$$h_e = \frac{\varphi}{2}\left(1 - \frac{G_\varkappa M_n}{\varrho RT}\right) \qquad (27)$$

The subscript e designates the extrapolation value. Recently Gottlieb [80] has derived an analytical expression for h as a function of $\varkappa$ for a perfect network. That was done unfortunately with the parameter p of the previous Flory theory [3] instead of the recently introduced parameter ζ [29]. Nevertheless, this expression

$$h = 1 - \varkappa^2(\varkappa^2 + 1)(\varkappa + 1)^{-4} \qquad (28)$$

can be used for $p = 2$ or $\zeta = 0$. A more general expression can be derived from:

$$h = 1 - \lim_{\lambda \to 1} \{[\lambda K(\lambda^2) - \lambda^{-2} K(\lambda^{-1})] (\lambda - \lambda^{-2})^{-1}\}$$

with the function K given by Eq. (17). This was also done by Gottlieb (see Appendix 1). The analytical expression of h derived by the present authors and reported in Appendix 1 is in agreement with this result. The values of h and h_e thus obtained are listed in the last columns of Table 4. They are quite similar. If h is compared to values given in literature, our method provides values of h which are dependent on the molecular weight and on the functionality of the network. Graessley found $h = 0$ in polybutadiene [45] and ethylene-propylene copolymers [81] and Macosko [19] $0.7 < h < 1$ in PDMS. They have assumed h independent of the molecular weight and used the branching theory and the Langley plot. We shall see in the next Section that these last methods are in need of some modifications.

6 Imperfect End-Linked Networks

6.1 Branching Theory

All the results presented to this point concerned networks assumed to be perfect. Imperfect networks can also be prepared by using a non-stoichiometric ratio r or, inadvertently, by an incomplete crosslinking process.

Macosko and Miller [65, 82, 83] have proposed a probability approach called the branching theory to derive the extent p of the reaction of the crosslinking agent, the number of elastically active strands v_a, the number of elastically active junctions μ_a from the value of the sol fraction w_s. This method was also extended to the calculation of other molecular parameters of gels and networks [84, 85]. The formulas of the branching theory have been rederived for the $(A_3 + B_2)$, $(A_4 + B_2)$ and $(A_f + B_2)$ systems, and are reported in Appendix 2. Macosko and Merrill generally study imperfect networks; they recognize that the crosslinking reaction is always incomplete and determine the extent of reaction from a corrected value of the sol fraction. They certainly overestimate the amount of unreacted chains because as pointed out above they neglect the formation and presence of cyclics in the sol fraction. But as no solution of this problem exists at present, further considerations are restricted to analysis of their results, which were recalculated and are reported in Tables 5 and 6. Some differences in the extent of reaction p can be found relative to the values initially reported by Macosko [19]. This is due to the use of an iterative method to derive p from w_s which is more accurate than the graphical one [54]. The quantity M_n is the number-average molecular weight of the primary end-reactive chains B_f, f_e is an average (or effective) functionality of the junctions, T_e is the so-called trapped entanglement factor first introduced by Langley [10]. The last quantity is the probability that all of the four directions from two randomly chosen points in the system, which may potentially contribute an entanglement, lead to the gel fraction.

Table 5. Structural parameters of imperfect trifunctional networks determined by the branching theory

$\frac{G}{vRT}$	Sources	M_n g mol^{-1}	T K	w_s	M_{A_f} g mol^{-1}	r	p	f_e	T_e	v_aRT N mm^{-2}	μ_aRT/T_e N mm^{-2}	G N mm^{-2}	$2C_1$ N mm^{-2}
1.72	Mark [52]	32900	298	0.030	190	1.00	0.889	2.67	0.467	0.0291	0.0415	0.0669	0.0342
1.88		25600		0.029			0.890	2.67	0.473	0.0379	0.0533	0.0947	0.0440
1.84		18500		0.030			0.889	2.67	0.467	0.0516	0.0738	0.1266	0.0660
0.87		9500		0.011			0.923	2.77	0.641	0.1426	0.1484	0.1496	0.0918
0.58		4700		0.030			0.889	2.67	0.465	0.2026	0.2903	0.1570	0.146
0.59		4000		0.021			0.902	2.71	0.532	0.2749	0.3445	0.2072	0.189
2.49	Meyers [68]	21600	298	0.0180	343	0.996	0.910	2.73	0.561	0.0542	0.0644	0.168	
1.77		15200		0.0129		1.01	0.911	2.73	0.616	0.0841	0.0910	0.203	
1.41		11100		0.0281		1.00	0.891	2.67	0.478	0.0884	0.1231	0.165	
1.25		8800		0.0173		1.00	0.908	2.72	0.567	0.1340	0.1576	0.210	
2.95	Macosko [54]	39900	297	0.103	330	0.99	0.841	2.52	0.212	0.0111	0.0347	0.0527	
2.39		30000	302	0.0325		1.09	0.835	2.51	0.451	0.0284	0.0420	0.112	
1.65		22400	285.5	0.073		0.998	0.853	2.56	0.283	0.0247	0.0582	0.0622	
1.14		11400	298.5	0.0073		1.11	0.863	2.59	0.700	0.1153	0.1098	0.193	
0.96			298.5	0.0060		1.21	0.814	2.44	0.729	0.1087	0.0995	0.192	
0.87			298.5	0.0040		1.31	0.775	2.32	0.781	0.1070	0.0914	0.221	
0.53		5430	297.5	0.119		0.986	0.835	2.51	0.182	0.0710	0.2593	0.062	
0.44		4190	290.5	0.0255		1.15	0.813	2.44	0.501	0.2061	0.2741	0.169	
0.53		3280	295.0	0.0091		1.10	0.864	2.59	0.671	0.3820	0.3796	0.296	

Table 6. Structural parameters of imperfect tetrafunctional networks determined by the branching theory

Sources	M_n g mol^{-1}	T K	w_s	M_{A_f} g mol^{-1}	r	p	f_e	T_e	$\nu_a RT$ N mm^{-2}	$\mu_a RT/T_e$ N mm^{-2}	G N mm^{-2}	$2C_1$ N mm^{-2}
Mark [6]	45000	298	0.075	208	1.00	0.788	3.15	0.278	0.0186	0.0411	0.0680	0.0383
	32900		0.046			0.822	3.29	0.380	0.0336	0.0534	0.0999	0.0579
	25600		0.017			0.881	3.52	0.571	0.0620	0.0629	0.1396	0.0843
	18500		0.06			0.804	3.21	0.324	0.0519	0.0976	0.1287	0.0892
	9500		0.03			0.849	3.39	0.466	0.1395	0.1779	0.2170	0.167
	4700		0.034			0.841	3.36	0.439	0.2678	0.3638	0.3843	0.353
	4000		0.012			0.897	3.59	0.625	0.4287	0.3923	0.4155	0.395
Llorente [61]	18500	298	0.075	208	1.00	0.788	3.15	0.277	0.0451	0.1001	0.1353	0.0925
Llorente [55]	11300	298	0.005	329	1.00	0.931	3.72	0.744	0.1750	0.1301	0.2786	0.196
			0.006			0.925	3.70	0.723	0.1710	0.1318	0.2840	0.169
			0.004			0.938	3.75	0.768	0.1794	0.1282	0.2749	0.199
			0.003		1.33	0.733	2.93	0.801	0.1522	0.1131	0.2748	0.183
			0.005			0.724	2.90	0.748	0.1432	0.1146	0.2756	0.178
			0.004			0.728	2.91	0.773	0.1474	0.1139	0.2850	0.165
Meyers [56,68]	21600	298	0.0049	402	1.121	0.845	3.40	0.800	0.0903	0.0644	0.240	0.142
	15200		0.0081		1.00	0.914	3.66	0.684	0.1216	0.1001	0.258	
	11100		0.0012		1.10	0.880	3.52	0.867	0.1903	0.1224	0.294	0.207
	8800		0.0049		1.01	0.923	3.69	0.745	0.2238	0.1668	0.328	0.244
Macosko [54]	30000	302	0.0242	329	1.20	0.742	2.97	0.508	0.0419	0.0501	0.146	
	22400	304	0.0161		1.11	0.807	3.20	0.580	0.0677	0.0692	0.176	
	11400	299.5	0.0025		1.10	0.868	3.47	0.813	0.1767	0.1228	0.291	
		298.5	0.0014		1.18	0.823	3.29	0.858	0.1757	0.1174	0.303	
		302	0.0040		1.17	0.812	3.25	0.769	0.1628	0.1229	0.295	
	4190	297	0.0041		1.20	0.794	3.17	0.768	0.4264	0.3250	0.461	

6.2 Structure Factors Obtained in Elongation Studies

6.2.1 High-Deformation Limit

For imperfect networks, the comparison of the ratio $2C_1/\nu RT$ with $1 - 2/\varphi$ is no longer possible because the number of effective chains ν is not known, and because not all of the junctions have the same functionality. Nevertheless, $2C_1$ can be compared with the value of the phantom modulus, which is $(\nu_a - \mu_a) RT$ [see Eq. (4)], and also with $(\nu - \mu) RT$. The variation of the ratio $2C_1/(\nu_a - \mu) RT$ with M_n is reported in Fig. 8 and 9 for trifunctional and tetrafunctional PDMS networks, respectively. Here M_n is different from M_c because an active chain can be formed by two or more chains bound with difunctional junctions. The constant $2C_1$ is always higher than the phantom modulus and the conclusions are similar to those reached in Section 4.2.

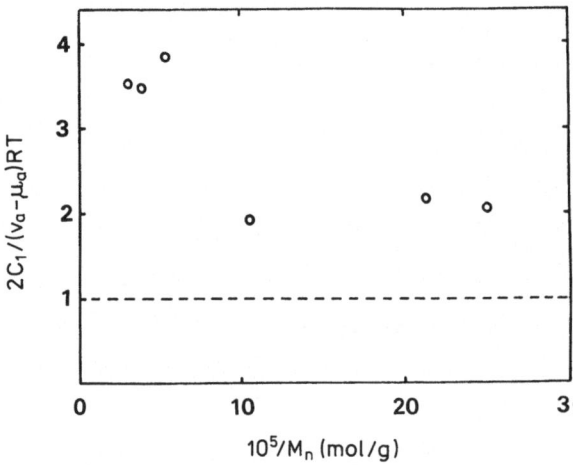

Fig. 8. Dependence of the ratio between $2C_1$ and the phantom modulus on the number-average molecular weight of primary chains for imperfect trifunctional end-linked PDMS networks; Mark [52] ($\bigcirc$)

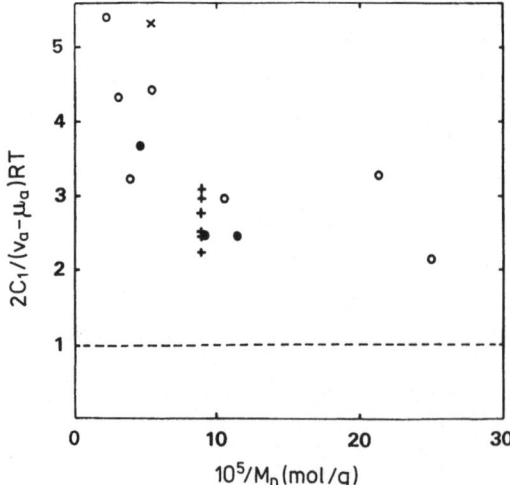

Fig. 9. Dependence of the ratio between $2C_1$ and the phantom modulus on the number-average molecular weight of primary chains for imperfect tetrafunctional end-linked networks: Mark [6] ($\bigcirc$); Llorente [61] ($\times$); Llorente [55] ($+$); Meyers [56,68] ($\bullet$)

6.2.2 Low-Deformation Limit

The branching theory in its latest refinement [19] furnishes the number of elastically active strands v_a. Now the affine modulus is vRT [Equation (10)] where v is the number of effective chains. The difference between v and v_a can be illustrated by following Flory's original idea [13]. When two free ends of the network are joined, the cycle rank ξ increases by one (see definition of ξ in Section 2.1). The corresponding increase in the number of effective chains is invariably $\Delta v = 2$. Following the chain leading from one of these free ends toward the network, we note the number of pathways i to the network from the next junction j offering two or more pathways to the network. The quantities i^1 and j^1 will be used for the other free end. The increase in the number of active chains Δv_a, according to the Scanlan-Case criteria, resulting from joining the pair of free ends is half the sum of the increases in the numbers of active chains attached to the two identified junctions j and j^1

$$\Delta v_a = (1/2) (v_{af,j} - v_{ai,j} + v_{af,j^1} - v_{ai,j^1}) \qquad (29)$$

where $v_{ai,j}$ and v_{ai,j^1} are the initial numbers of active chains for the junctions j and j^1, respectively and $v_{af,j}$ and v_{f,j^1} are the final numbers. The increase Δv can be calculated for the three examples

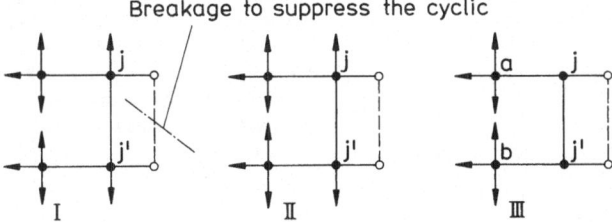

Breakage to suppress the cyclic

Where junctions are shown by filled circles, chain ends by open circles, and arrows identify chains leading to the network. The results are

I		II		III	
j	j^1	j	j^1	j	j^1
$i = 3$	$i^1 = 3$	$i = 3$	$i^1 = 2$	$i = 2$	$i^1 = 2$
$v_{ai,j} = 3$	$v_{ai,j^1} = 3$	$v_{ai,j} = 3$	$v_{ai,j^1} = 0$	$v_{ai,j} = 0$	$v_{ai,j^1} = 0$
$v_{af,j} = 4$	$v_{af,j^1} = 4$	$v_{af,j} = 4$	$v_{af,j^1} = 3$	$v_{af,j} = 3$	$v_{af,j^1} = 3$
$\Delta v_a = 1$		$\Delta v_a = 2$		$\Delta v_a = 3$	

Thus Δv_a depends on the topology of the network near the joining of the free ends while Δv is independent. As this difficulty can't be solved easily, we assume in a first approximation that the networks are not too imperfect and so that v_a is not far from

v. We shall see later a possible way to obtain v. Then we can use the ratio G/v_aRT. The structure factors reported in Tables 5 and 6 are shown in Fig. 10 and 11 for trifunctional and tetrafunctional end-linked PDMS networks, respectively. The comparison of Fig. 3 and 10 shows that even if the crosslinking process is supposed incomplete the structure factor is less than one for $M_n < 10^4 \, g \, mol^{-1}$. Therefore it is still not necessary to assume an entanglement contribution. In the region of $M_n > 10^4 \, g \, mol^{-1}$ the values of A'_4 in Fig. 11 are around unity and may be less if the sol fraction is corrected by accounting for cyclics. Again in the region $M_n > 10^4$ the structure factors are higher than unity but the highest values (4–5) are certainly overestimated. Thus the branching theory leads to conclusions not very different from those already derived by assuming perfect networks (Sect. 4.2).

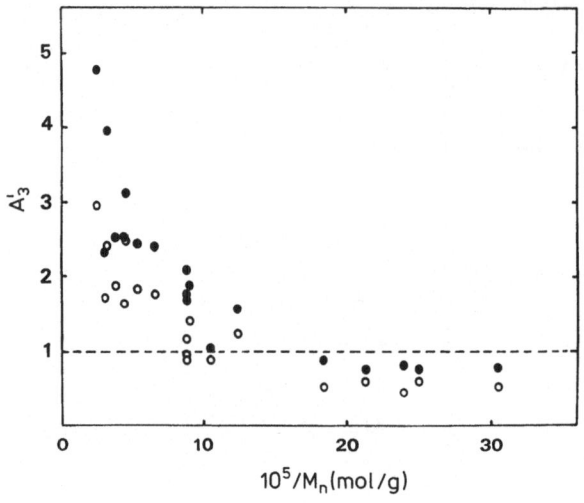

Fig. 10. Dependence of the structure factor at zero deformation on M_n for imperfect trifunctional end-linked PDMS networks. Open circles correspond to $A'_3 = G/vRT$ and filled circles to $A'_3 = G/v_aRT$

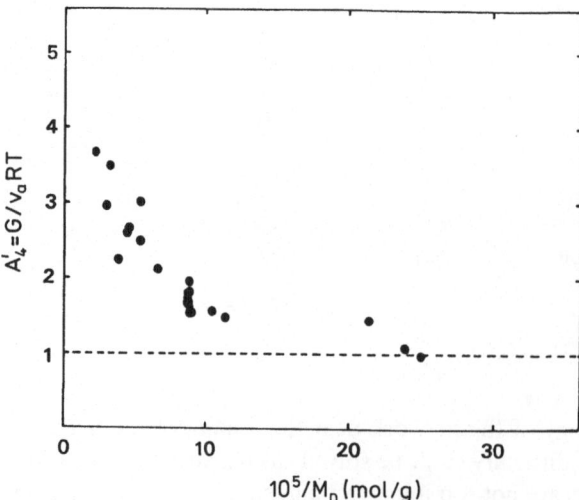

Fig. 11. Dependence of the structure factor at zero deformation on M_n for imperfect tetrafunctional networks

6.3 The Flory Theory

We saw in the previous paragraph that the structure factor at zero deformation is less than unity for imperfect trifunctional networks when $M_n < 10^4$ g mol^{-1}. For these low molecular weights between crosslinks, the non-equilibrium effects are certainly avoided and it is then possible to apply the Flory theory. The value of the phantom modulus is given by the branching theory; however it is necessary to calculate the factor $\dfrac{\mu}{\xi}$ of Eq. (16), which is the ratio of the number of effective junctions to the cycle rank. For a perfect network of functionality φ

$$\frac{\mu}{\xi} = \frac{2}{\varphi - 2} \tag{30}$$

This relationship comes from Eq. (7) with

$$\mu = \mu_a = 2 v_a / \varphi$$

The problem of counting the number of effective chains and junctions in an imperfect network has not yet been solved. In the case of end-linked networks, two points of view can be considered.

The usual one is to take into account only the active chains and junctions. The difunctional junctions are thus not included. Therefore the molecular weight between crosslinks can be different from M_n and its distribution is unknown. Nevertheless one gains more knowledge about the functionality of the considered junctions, three in an imperfect trifunctional network, a mixture of three and four in a tetrafunctional one. The Flory parameter $\varkappa$ and the Graessley parameter h are not defined. One can take an average over the distribution of molecular weights between crosslinks and the different functionalities.

Another point of view is to include the difunctional junctions; we shall assign the name effective chains and junctions to the chains and junctions thus counted. This procedure is a little artificial since the difunctional junctions are not constrained. The functionality of the network is less defined than in the preceding case, but now the molecular weight between crosslinks is M_n.

In the following development, the difference in the interpretation of data according to the two points of view will be shown. We can already emphasize that the deliberate use of imperfect networks leads to unknown structures and the use of one of the two concepts is only approximate. If we consider the crosslinking process of A_f molecules (i.e., with f reactive groups) with B_2 oligomers, at a given extent of reaction p of the A groups, the total number of junctions is

$$\mu_{tot} = \sum_{i=0}^{f} \mu_i \tag{32}$$

where μ_{tot} is also the initial number of crosslinking molecules introduced, and μ_0 is the number of junctions which are still in the sol. The quantity μ_1 is the number of junctions having only one group reacted and thus these junctions are at the ends of dangling chains. We notice that the end of a dangling chain can also be free of cross-linking agent and be only a non-reacted oligomer function. Then μ_i is the number of junctions when i groups have already reacted. The number of effective junctions is by definition

$$\mu = \sum_{i=2}^{f} \mu_i \tag{33}$$

and the number of elastically active junctions is

$$\mu_a = \sum_{i=3}^{f} \mu_i \tag{34}$$

It follows that

$$\mu = \mu_2 + \mu_a \tag{35}$$

The total number g_0 of reactive crosslinking A groups at the beginning of the reaction is

$$g_0 = f\mu_{tot} \tag{36}$$

Here all the molecules are assumed to have the same number f of reactive groups. The total number of groups which have reacted at the extent of reaction p is

$$g = pg_0 = pf\mu_{tot} \tag{37}$$

But a junction of functionality i in the network has i reacted groups; therefore g is also

$$g = \sum_{i=1}^{f} i\mu_i \tag{38}$$

The combination of Eq. (32), (37) and (38) leads to

$$pf \sum_{i=0}^{f} \mu_i = \sum_{i=1}^{f} i\mu_i \tag{39}$$

For p far higher than p_{gel}, μ_0 and μ_1 may be neglected; Eq. (39) then becomes

$$pf\mu = \sum_{i=2}^{f} i\mu_i \tag{40}$$

This assumption means that the probability to find at $p \geq 0.9$ a molecule with f unreacted groups or $(f-1)$ unreacted groups is small. The crosslinking process for PDMS is done generally at temperatures far above T_g and so the high mobility acts in favor of this assumption. Equation (40) can be solved without any further assumptions only when f = 3 and by using Eq. (34) and (35). Specifically,

$$\mu = \frac{\mu_a}{3p - 2} \tag{41}$$

It's obvious that in high functionality networks v_a is not very different from v. The number of difunctional junctions μ_2 is expected to be small relative to the number of junctions of higher functionalities if the reaction is near completion. It follows from Eq. (2), (3) and (41) that

$$\frac{\mu}{\xi} = \frac{\mu_a}{(3p - 2)(v_a - \mu_a)}$$

For a trifunctional network, the value of $\frac{v_a}{\mu_a}$ given by the branching theory is simply $\frac{3}{2}$. Hence

$$\frac{\mu}{\xi} = \frac{2}{3p - 2} \tag{42}$$

instead of 2 for perfect networks. To proceed further with this idea, we can now derive the number of effective chains for a trifunctional network for $v - \mu = v_a - \mu_a$ [Eq. (3)] and $\frac{v_a}{\mu_a} = \frac{3}{2}$:

$$v = v_a \left(\frac{p}{3p - 2}\right) \tag{43}$$

(As expected, $v = v_a$ when $p = 1$). For a trifunctional imperfect network, v is always larger than v_a. That is quite understandable because the network corresponds mostly to Scheme III of Section 6.1 (the junctions a and b are now trifunctional). In this case, when the number of effective chains increases by two, the number of active chains increases by three, and joining two free ends means increasing p ($p \leq 1$). Therefore to have $v = v_a$ at $p = 1$ (perfect network) it is necessary that v_a is less than v for $p < 1$. The fitting of the experimental reduced forces versus the inverse of deformation by the Flory theory is shown in Fig. 12 for imperfect trifunctional networks of $M_n = 4000$, 4700 and 9500 g mol^{-1}. The experimental data are due to Mark [52]. The values of the phantom modulus used as starting points of the theoretical curves are those given in Table 5, and the same method of fitting as described earlier was employed. It is obvious from the curves that ζ must be chosen equal to zero. The results are given in Table 3 for the value $\frac{\mu}{\xi} = 2$, assuming a perfect network, and for the value calculated by equation (42). For $M_n = 4000$ and 4700 g mol^{-1}, the error range in fitting is large, and values of the phantom modulus calculated by the branching theory are too small. That is not surprising because these networks of low molecular weights are certainly nearly perfect but the branching theory predicts from the sol

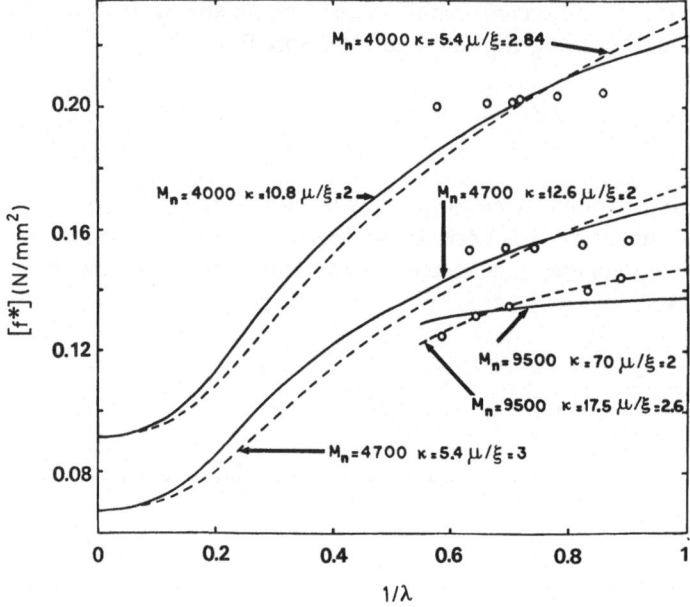

Fig. 12. Dependence of the parameters of the Flory theory for imperfect trifunctional networks ($\zeta = 0$). The results show the influence of the ratio μ/ζ

fraction extents of reaction p = 0.902 for M_n = 4000 and p = 0.889 for M_n = 4700, which are the same as for M_n = 32 900, which is highly doubtful. The results obtained with M_n = 9500 g mol^{-1} are better; the branching theory predicts p = 0.923 and a good fit is obtained with $\frac{\mu}{\xi}$ derived from Eq. (42). The value of $\varkappa$ is reduced (from 70 to 17.5) when $\frac{\mu}{\xi}$ is changed (from 2 to 2.8) and obviously is different from $\varkappa$ = 3.4 obtained previously for the same network when it was assumed to be perfect. That is interesting because the expected range in $\varkappa$ can be determined (from 3.4 to 17.5).

Table 7. Values of the Flory parameter $\varkappa$ for imperfect trifunctional networks ($\zeta = 0$), using experimental data from Ref. 52

M_n g mol^{-1}	p	$\dfrac{\mu}{\zeta}$	$\varkappa$	$\varkappa f_{ph}^{1/2}$	h[a]	h_e[b]
4000	0.902	2	10.8	3.3		
4000	0.902	2.84	5.4	1.6	0.56	0.47
4700	0.889	2	12.6	3.3		
4700	0.889	3	5.4	1.4	0.56	0.47
9500	0.923	2	70	15.3		
9500	0.923	2.6	17.5	3.8	0.19	0.20

[a] Determined from Fig. 15;
[b] determined from Fig. 12

Again, it is not possible to find a constant value of $\varkappa f_{ph}^{1/2}$ (Table 7). We can now derive the real structure factor at zero deformation for trifunctional networks

$$A_3' = G/\nu RT$$

the values of which were reported in Table V and are shown in Fig. 10 (open circles). The use of the number of effective chains reduces A_3' to reasonable values but doesn't change the conclusions already reached in the preceding Section.

6.4 The Langley Plot

6.4.1 General Comments

The main phenomenological approach for rubber elasticity at small deformations is due to Langley [10,44]. The shear modulus is assumed to be the sum of two terms

$$G = G_c + G_e T_e \tag{44}$$

where G_c is the contribution of the chemical crosslinks. Taking into account the restrictions of the fluctuations of junctions as in the Flory theory leads to the previously presented Eq.

$$G_c = \left(1 - h\,\frac{\mu_a}{\nu_a}\right) \nu_a RT \tag{45}$$

This expression derived from Eq. (1) is rigorously valid only for perfect networks. In a real network

$$G_c = \left(1 - h\,\frac{\mu}{\nu}\right) \nu RT \tag{46}$$

A discussion of this subject is presented below. The additional contribution $G_e T_e$ is said to arise from permanently trapped (inter-chain) entanglements in the network. Linear polymers of high molecular weight exhibit a storage modulus $G'(\omega)$ which remains relatively constant over a wide range of frequencies [78]. This plateau modulus G_N^0 is independent of chain length for long chains and is insentitive to temperature. Since it varies as the square of the volume fraction of polymer in concentrated solutions, it could be due to pair-wise interactions between chains [86]. Recently Graessley and Edwards have proposed a universal law for the dependence of G_N^0 on the chemical structure of polymers [87]. During the crosslinking process a portion of such interactions or entanglements could be trapped in the network and act then like physical junctions. Ferry has tried to prove this statement by irradiation crosslinking of already deformed networks [88]. The factor T_e is the probability that the chains so entangled are elastically active. The modulus G_e is thought to have a value close to G_N^0. Hence, the small-strain modulus becomes

$$G = \left(1 - h\,\frac{\mu_a}{\nu_a}\right) \nu_a RT + G_e T_e \tag{47}$$

For a real network, v_a, μ_a and T_e are determined by the branching theory, and G is the experimental value of the modulus. It is worth noting the point of view of Flory on this question of entanglements [28]. According to his analysis, an entanglement can not be equivalent to a chemical crosslinkage. Contacts between a pair of entangled chains are transitory and of short duration owing to the diffusion of segments and associated time-dependent changes of configuration. Such entanglements as previously described are considered to be of minor importance in equilibrium stress measurements.

Equation (47) can be rewritten

$$\frac{G}{T_e} = \left(1 - h\,\frac{\mu_a}{v_a}\right)\frac{v_a RT}{T_e} + G_e \tag{48}$$

The corresponding plot of G/T_e versus $v_a RT/T_e$ is called the Langley plot and has been widely used [19]. Experimental points representing imperfect networks of the same initial functionality of the crosslinking agent but different molecular weights of the primary chains are generally fitted by a straight line of slope $\left(1 - h\,\frac{\mu_a}{v_a}\right)$ and of value G_e at the origin of the x-axis. The quantity $\frac{\mu_a}{v_a}$ is assumed to be nearly constant so h can be determined from the slope. G_e is frequently found to be close to G_N^0 which has been taken to suggest the existence of an entanglement contribution. It is necessary to emphasize that such a fitting is based on two doubtful assumptions. The first is that the restrictions of fluctuations of junctions taken into account by the parameter h are independent of the molecular weight of the primary chains. This is in complete disagreement with the Flory theory and the results previously obtained presented in this review. The second is that the maximum contribution of entanglements G_e is also independent of the molecular weight. What is true for uncrosslinked polymers of $M_n \geqq 10^5$ g mol^{-1} is certainly invalid for the networks studied, whose range of M_n is 3000–50,000 g mol^{-1}. Moreover the contribution of an entanglement obtained by dynamic measurements at $\omega \sim 1000$ rad/s could be quite different from the one obtained at equilibrium owing to possible conformational rearrangements of the chains. Finally each experimental sample i of the Langley plot presents a topological entity and Eq. (47) is then

$$G^{(i)} = \left(1 - h^{(i)}\,\frac{\mu_a^{(i)}}{v_a^{(i)}}\right) v_a^{(i)} RT + G_e^{(i)} T_e^{(i)} \tag{49}$$

A last comment on G_e deals with the comparison of two networks: one of small M_n and the other of large M_n. The distance between entanglements is known for a given polymer [78]. If the molecular weight increases between two crosslinks, that means the increase of the number of entanglements between these two points. Therefore in high molecular weight polymers the entanglements will be mostly surrounded by other entanglements while in small molecular weight polymers they will be mostly surrounded by crosslinks. This is an extreme picture for PDMS networks because the distance between entanglements is large (8100 g mol^{-1}) and so it can be only one or two entanglements per chain. But it holds for polybutadiene (~ 3000 g mol^{-1}) whose non-equilibrium important entanglement contribution is well-known [89].

6.4.2 A Modification and Its Application

We propose to use another form of Eq. (47)

$$(G - \nu_a RT)/T_e = G_e - h\mu_a RT/T_e \qquad (50)$$

The suggested plot of $(G - \nu_a RT)/T_e$ versus $\mu_a RT/T_e$ has seldom been done [38, 77]. The numerical values of Tables 5 and 6 are represented in Fig. 13 for imperfect trifunctional networks and in Fig. 14 for tetrafunctional ones. It is necessary to comment in some detail on this type of representation. Each point belongs to a straight

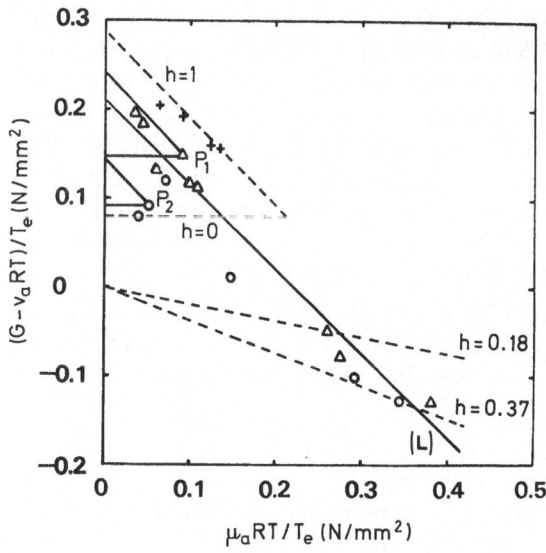

Fig. 13. Plot of Eq. (50) for imperfect trifunctional PDMS networks: Mark [52] ($\bigcirc$); Meyers [68] ($+$); Macosco [54] ($\triangle$)

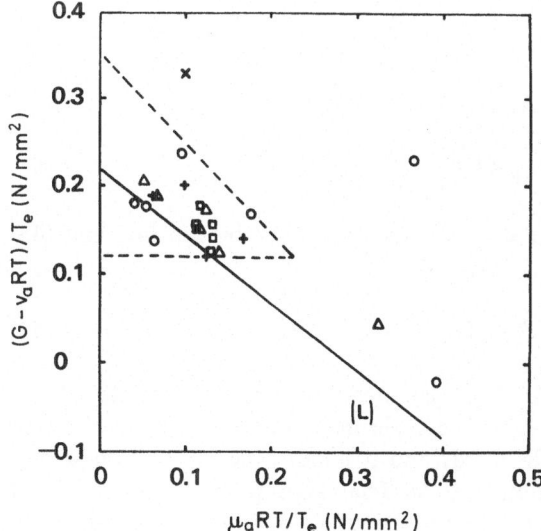

Fig. 14. Plot of Eq. (50) for imperfect tetrafunctional PDMS networks: Mark [6] ($\bigcirc$); Meyers [56,68] ($+$); Macosko [54] ($\triangle$); Llorente [55] ($\square$); Llorente [61] ($\times$)

line of slope −h (with $0 \leq h \leq 1$) which cuts the vertical coordinate axis at a value $G_e \geqq 0$. These restrictions directly come from the physical meaning of the different terms of Eq. (50) without any further assumption. All the points situated below the $\mu_a RT/T_e$ axis of Fig. 13 and corresponding to the lowest M_n ($3280 \leq M_n \leq 5430$ g mol^{-1}) of Table 5 can belong to lines passing through the origin (i.e., $G_e = 0$) and with h between 0.18 and 0.37. Thus, one doesn't need to assume any interaction or entanglement contribution at elastic equilibrium. The stress curve is fitted by the Flory theory as shown in Fig. 12. Also the point for $M_n = 9500$ g mol^{-1} which lies just above the x-axis of Fig. 13 has a value of $(G - v_a RT)/T_e$ which is slightly over-estimated by using $(2C_1 + 2C_2)$ instead of the real small-strain modulus and must be below it else the fitting of Fig. 12 could not be possible. For the other points it is necessary to have $G_e \neq 0$ or $G_e > 0$, which would be consistent with an entanglement contribution to the modulus at small deformations. It is also consistent, of course with failure to reach elastic equilibrium, or an overestimated value of the sol fraction, or both. The point P_1 may have a parameter h between 0 and 1 and a parameter G_e between 0.15 and 0.24 N mm^{-2}; for the point P_2, $0 \leq h \leq 1$ and $0.09 \leq G_e \leq 0.14$ N mm^{-2}. That is in disagreement with Gottlieb et al. [19] who predict all the points to be on the straight line (L) (h = 0.95 and $G_e = 0.21$ N mm^{-2}). The dotted triangle determines the range of acceptable values of G_e if all the points are used (0.08–0.29 N mm^{-2}). The value of G_N^0 for PDMS (0.24 N mm^{-2}) is included in this interval. The same considerations hold for the tetrafunctional networks of Fig. 14. Because of experimental error in G, the points just above the x-axis could actually be below. The line (L) of Gottlieb et al. has the parameters h = 0.765 and $G_e = 0.22$ N mm^{-2}.

6.4.3 Effective Chains and Junctions

Before proceeding in this review, we treat now in more detail the difference between the use of v_a and μ_a instead of v and μ. Equation (46) tells one that

$$G_c = \left(1 - h \frac{\mu}{v}\right) vRT$$

and hence Eq. (50) becomes

$$(G - vRT)/T_e = G_e - h\mu RT/T_e \tag{51}$$

The quantities v and μ are given for a trifunctional network by Eq. (41) and (43) and permit one to write Eq. (51) as

$$(G - v_a RTp/(3p - 2))/T_e = G_e - h\mu_a RT/((3p - 2) T_e) \tag{52}$$

The values of $\mu RT/T_e$ and $(G - vRT)/T_e$ calculated from the data of Table 5 are listed in Table 8 and reported in Fig. 15. The experimental points of networks with $M_n < 10^4$ g mol^{-1} are now below the x-axis and it is not necessay to assume any interaction or entanglement contribution for them. That is equivalent to the conclusion reached on Fig. 10 (open cycles) because instead of comparing G/vRT to unity

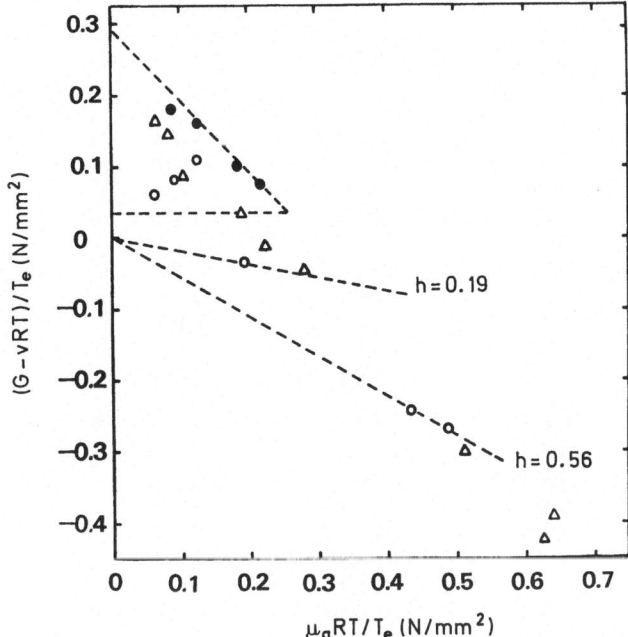

Fig. 15. Plot of Eq. (51) for imperfect trifunctional networks: Mark [52] (○); Meyers [68] (●); Macosko [54] (△)

we compare $(G - \nu RT)$ to zero. Assuming $G_e = 0$, we can obtain the parameter h. Also a similar extrapolation of the Flory curves of Fig. 12 at $\lambda^{-1} = 1$ can be done as in Section 5.2. One can easily derive the expression

$$h_e = p\left(\frac{\nu_a}{\mu_a}\right)(1 - G_x(3p - 2)/(p\nu_a RT)) \tag{53}$$

In an imperfect trifunctional network, $\dfrac{\nu_a}{\mu_a} = \dfrac{3}{2}$. The values of h_e are listed in Table 8, and are obviously in agreement with h when the fitting is good, i.e. when the branching theory doesn't underestimate the extent of reaction. The analytical value of h as a function of x can be derived for an imperfect network from Eq. (46) with the Flory reduced force at $\lambda^{-1} = 1$:

$$\nu - h\mu = \lim_{\lambda \to 1} \{\xi + \mu[\lambda K(\lambda^2) - \lambda^{-2}K(\lambda^{-1})](\lambda - \lambda^{-2})^{-1}\}$$

Using Eq. (2), this leads to

$$h = 1 - \lim_{\lambda \to 1} \{[\lambda K(\lambda^2) - \lambda^{-2}K(\lambda^{-1})](\lambda - \lambda^{-2})^{-1}\} \tag{54}$$

One can also derive a more complicated form from Equation (45).

Table 8. Values of the parameters of Eq. (51) for imperfect trifunctional networks

Sources	M_n g mol^{-1}	p	$\mu RT/T_e$	$(G - \nu RT)/T_e$
Mark [52]	32900	0.889	0.0622	0.0604
	25600	0.890	0.0796	0.0939
	18500	0.889	0.1106	0.1237
	9500	0.923	0.1930	—0.0338
	4700	0.889	0.4352	—0.2427
	4000	0.902	0.4880	—0.2707
Meyers [68]	21600	0.910	0.0882	0.1789
	15200	0.911	0.1241	0.1599
	11100	0.891	0.1829	0.1007
	8800	0.908	0.2177	0.0739
Macosko [54]	39900	0.841	0.0663	0.1648
	30000	0.835	0.0832	0.1442
	22400	0.853	0.1041	0.0864
	11400	0.863	0.1864	0.0344
		0.814	0.2251	—0.0116
		0.775	0.2812	—0.0440
	5430	0.835	0.5135	—0.3024
	4190	0.813	0.6244	—0.4242
	3280	0.864	0.6412	—0.3897

6.4.4 High-Functionality Networks

The simplified equations of the branching theory for this type of network have been given explicitly by Gottlieb et al. [19]. Here one may assume that the numbers of active chains and junctions are respectively equal to the numbers of effective ones (see Section 6.2). A very interesting study of these imperfect systems was done by Merrill [56, 59, 90, 91]. The structural parameters of end-linked PDMS networks of initial functionalities of the crosslinking agents f = 24 and 44 are reported [19]. We have checked these values and calculated the parameters $\mu_a RT/T_e$ and $(G - \nu_a RT)/T_e$ of the modified Langley plot and also the effective functionality f_e (Table 9) defined as:

$$f_e = pf \tag{55}$$

This effective functionality varies from 12 to 32 and the molecular weight of the primary chains from 8800 to 51000 g mol^{-1}. Thus these results permit the analysis of the dependence of the modified Langley plot on chain length and junction functionality. The data are reported in Fig. 16. To include in the same graph the influence of M_n and f_e, we have used circles for the lowest molecular weights and squares for the highest ones with an arbitrary cutoff at $M_n = 17000$ g mol^{-1} to get the clearest presentation. Likewise, open points represent the lowest effective functionalities and filled points the highest ones. The points with a star must be omitted from the general weighting since they correspond to mixed conditions (large molecular weight with very low functionality or highest functionality at constant M_n too low compared with the highest values). All the values of $(G - \nu_a RT)$ are positive. Llorente [55] also obtained structure factors at zero deformation greater than unity ($A'_\varphi \sim 1.40$) (see Table 3). Therefore it seems that for these systems another contribution should be

Table 9. Imperfect high functionality PDMS networks

M_{A_f} g mol^{-1}	f	M_n g mol^{-1}	T K	w_s	r	f_e	$\mu_a RT/T_e$ N mm^{-1}	$(G - \nu_a RT)/T_e$ N mm^{-2}	T_e
1602	24	8800	298	0.0015	1.20	19.2	0.030	0.337	0.85
		9320		0.0009	1.21	19.2	0.028	0.333	0.88
		10100		0.0020	1.20	19.1	0.027	0.308	0.83
		11100		0.0015	1.20	19.2	0.024	0.283	0.85
		12900		0.0028	1.20	18.9	0.022	0.273	0.80
		17000		0.0027	1.20	19.0	0.016	0.252	0.80
		21600		0.0022	1.21	18.9	0.013	0.262	0.82
		28100		0.0012	1.16	20.0	0.009	0.252	0.86
		12200		0.0018	1.26	18.2	0.023	0.267	0.84
		16300		0.0103	1.30	16.6	0.022	0.363	0.65
		20300		0.0294	1.65	12.0	0.029	0.313	0.49
		26100		0.0146	1.65	12.8	0.019	0.191	0.59
		35400		0.0148	1.66	12.7	0.014	0.242	0.59
		51000		0.0076	1.42	15.4	0.007	0.248	0.69
2802	44	8800		0.0017	1.11	38.0	0.016	0.288	0.84
		10100		0.0017	1.27	33.2	0.015	0.280	0.84
		11100		0.0045	1.20	34.2	0.014	0.242	0.75
		12900		0.0021	1.30	32.3	0.013	0.247	0.82
		17000		0.0019	1.31	32.1	0.009	0.241	0.83
		21600		0.0038	1.36	30.4	0.008	0.241	0.77
		28100		0.0033	1.45	28.6	0.007	0.264	0.78
		12200		0.0037	1.30	31.8	0.014	0.293	0.77
		16300		0.0121	1.81	21.6	0.017	0.341	0.62
		20300		0.0228	1.80	20.7	0.015	0.322	0.51
		26100		0.0120	1.80	21.8	0.011	0.250	0.62
		35400		0.0175	1.80	21.1	0.008	0.272	0.56
		51000		0.0105	1.51	26.1	0.005	0.235	0.64

added to that from the chemical crosslinks. Nevertheless high-functionality networks can not be included in such an analysis, as has been pointed out by Flory and Erman [92]. Further comments are useful with regard to Fig. 16. The filled points and the open squares are concentrated in the same area of the plot. The data with $f_e \simeq 19$ (open points) and $f_e > 30$ (filled points) are represented in Fig. 17. When the molecular weight increases the points for $f_e \simeq 19$ approach this area. Also the points for $f_e > 30$ tend to leave this area when the molecular weight decreases. The networks of very high functionality or very high molecular weight should have their junction fluctuations nearly suppressed so h can be assumed to be zero. A line of zero slope (dotted line in Fig. 16) fits all the points corresponding to these networks and leads to a value of $G_e = 0.25$ N mm^{-2}, which is close to G_N^0. Gottlieb et al. obtained h = 0 and $G_e = 0.24$ N mm^{-2} by using the Langley plot, fitting all the data with the same line (see Fig. 7 [19]). On the contrary, the modified Langley plot shows that the points of lowest functionality or molecular weight can not be fitted with the same line. For these data G_e will be more than 0.25, which doesn't mean that the entanglement contribution $G_e T_e$ will be higher since T_e also varies (see Table 9). In any case, it is not possible to derive the parameters h and G_e for those systems by knowledge only of the small strain modulus.

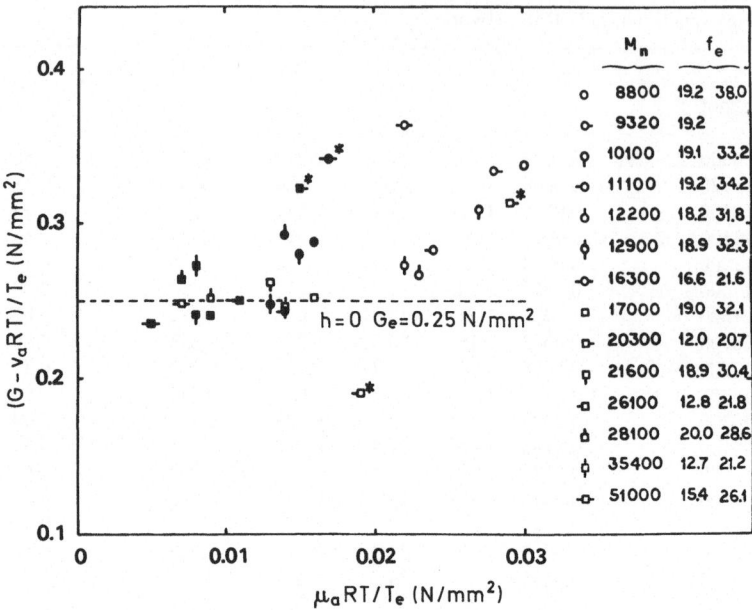

Fig. 16. Plot of Eq. (50) for imperfect high-functionality PMDS networks to show the influence of M_n and f_e. Open circles: lowest f_e at constant M_n; filled circles: highest f_e

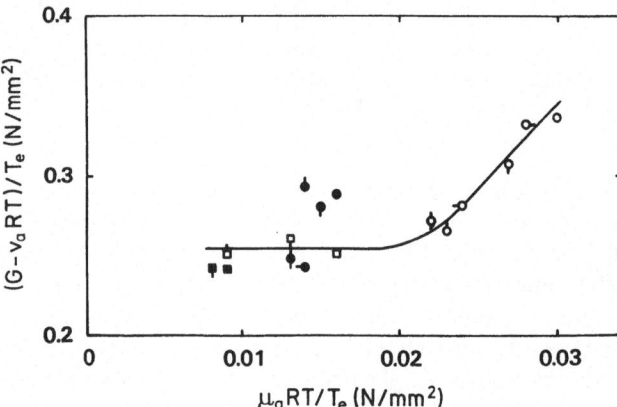

Fig. 17. Plot of Eq. (50) for high-functionality networks showing the influence of M_n (same symbols as in Fig. 16). Open points: $f_e \simeq 19$; filled points: $f_e > 30$

7 Conclusions

The aim of this paper is to clarify current discussions about rubber elasticity particularly with regard to entanglement contributions. [A similar review convering some of the same topics has recently been prepared by Eichinger [93)]. After a brief review of the existing molecular theories, it is shown that the Flory parameters can be determined by a new method of fitting the stress-strain curves obtained for perfect end-linked PDMS networks. For networks assumed to be imperfect, the branching theory is employed with the use of effective chains and junctions instead of active ones. A modified Langley plot is also proposed. Although the analysis of swelling results seems very promising [94], the scattering of data found in the literature proves that experimental limits have now been reached in the preparation and characterization of model networks. It is time to look for other independent quantitative information such as that provided by GPC, NMR and neutron scattering [95–99].

8 Appendices

8.1 Appendix 1

As Gottlieb and Macosko have done [80] the present authors have derived the relation between h and $\varkappa$ but with the parameter ζ of the latest Flory theory [29]. The result, also obtained by Gottlieb [100], is

$$h = 1 - \frac{1}{4}(2 - \zeta)^2 \left(1 + \frac{1}{\varkappa^2}\right)\left(1 + \frac{1}{\varkappa}\right)^{-4} \tag{61}$$

which gives the expected result [80] for $p = 2$ and $\zeta = 0$. The relation between $\varkappa$ and h is represented in Fig. 18 for different values of ζ. Whatever ζ is, h tends now to unity when $\varkappa$ tends to zero. But for $\varkappa \to \infty$, h reaches an asymptotic value (dashed line) which is zero only when $\zeta = 0$.

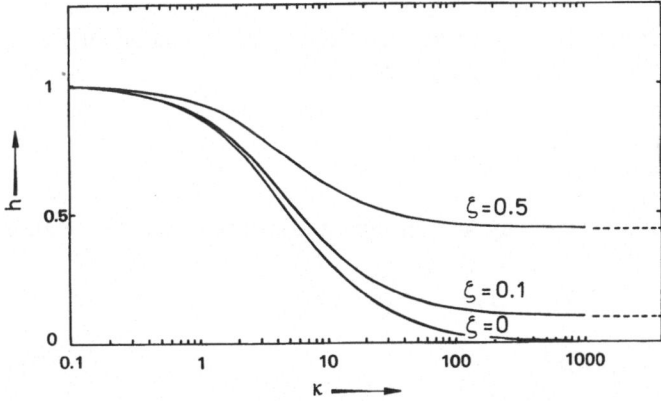

Fig. 18. Relation between $\varkappa$ and h for different values of ζ [Eq. (61)]. The dashed lines represent extrapolation of h to infinite $\varkappa$

8.2 Appendix 2

This Section is concerned with formulas to obtain p, v_a, μ_a, T_e and f_e from the sol fraction w_s by the branching theory for end-linked networks.

The networks are formed by joining primary end-reactive chains B_2 of number-average molecular weight M_n with crosslinking agent A_f of functionality f.

For the $(A_f + B_2)$ system, the sol fraction is given by the expression (Eq. (39) [83])

$$w_s = w_{A_f}[P(F_A^{out})]^f + w_{B_2}[P(F_B^{out})]^2 \tag{62}$$

where w_{A_f} is the weight fraction of crosslinking agent and is given by

$$w_{A_f} = (2/f)\, rM_{A_f}/((2/f)\, rM_{A_f} + M_n) \tag{63}$$

The quantity M_{A_f} is the molecular weight of crosslinking agent, and r is the initial ratio of reactive functions

$$r = \frac{f}{2}\frac{[A_f]_0}{[B_2]_0} \tag{64}$$

where $[A_f]_0$ and $[B_2]_0$ are the initial molar concentrations of the crosslinking molecules and the end-reactive oligomers, respectively. Specifically,

$$[A_f]_0 \simeq 2r\varrho/fM_n \tag{65}$$

where ϱ is the density of the polymer, and w_{B_2}, the weight fraction of reactive chains, is given by

$$w_{B_2} = M_n/[(2/f)\, rM_{A_f} + M_n] \tag{66}$$

The quantity $P(F_A^{out})$ is the probability that looking "out" from a functional group leads to a finite chain [65 b)] and is given by Eq. (25) [83]:

$$rp^2 \sum_{i=0}^{f-2} P(F_A^{out})^i - 1 = 0 \tag{67}$$

The quantity p is the extent of reaction in the A groups, and rp is the extent of reaction in the B groups [Eq. (12) [82']].

For an $(A_3 + B_2)$ system [Eq. (15) [83']]

$$P(F_A^{out}) = (1 - rp^2)/rp^2 \tag{68}$$

For an $(A_4 + B_2)$ system [Eq. (16) [83']]

$$P(F_A^{out}) = (1/rp^2 - 3/4)^{1/2} - 1/2 \tag{69}$$

The probability $P(F_B^{out})$ is obtained by combination of equations (11) and (12) [83]:

$$P(F_B^{out}) = 1 - (1/p)\,[1 - P(F_A^{out})] \tag{70}$$

The trapped entanglement factor T_e comes from Eq. (53) [83]

$$T_e = (1 - P(F_D^{in}))^4 \tag{71}$$

For long B_2's reacting with small A_f's the B_2's will be entangled and

$$P(F_D^{in}) = P(F_B^{out})$$

Eq. (A 8) [65a] is thus obtained:

$$T_e = (1 - P(F_B^{out}))^4 \tag{72}$$

The branching theory enables the calculation of the number of active junctions μ_a and chains ν_a (Eq. (A 6) and (A 7) [65a]):

$$\mu_a = [A_f]_0 \sum_{i=3}^{f} P(X_i, f) \tag{73}$$

$$\nu_a = [A_f]_0 \sum_{i=3}^{f} (i/2)\, P(X_i, f) \tag{74}$$

where (Eq. (45) [83])

$$P(X_i, f) = \frac{f!}{i!\,(f-i)!} \, [P(F_A^{out})]^{f-i} \, [1 - P(F_A^{out})]^i \tag{75}$$

For an $(A_3 + B_2)$ system:

$$\mu_a = (2r\varrho/3M_n)\, ((2rp^2 - 1)/rp^2)^3 \tag{76}$$

$$\nu_a = (r\varrho/M_n)\, ((2rp^2 - 1)/rp^2)^3 \tag{77}$$

and thus $\mu_a = 2\nu_a/3$.
   For an $(A_4 + B_2)$ system

$$\mu_a = (r\varrho/2M_n)\, [1 - P(F_A^{out})]^3 \, [1 + 3P(F_A^{out})] \tag{78}$$

$$\nu_a = (r\varrho/M_n)\, [1 - P(F_A^{out})]^3 \, [1 + 2P(F_A^{out})] \tag{79}$$

The effective functionality f_e is defined by [Eq. (A 5) [19]]

$$f_e = pf \tag{80}$$

8.3 Appendix 3

A calculation by Flory [33] of the degree of interpenetration of chains and junctions leads to

$$\varkappa = I\langle r^2\rangle_0^{3/2}\, (\mu/V_0)$$

where I is a constant of proportionality. The right-hand side of the equation is simply proportional to the number of junctions in the volume pervaded by a chain. It can then be shown that $\varkappa$ depends on M_n to the power 0.5. From our results reported in Table 4 an easy way to obtain the dependence of $\varkappa$ on M_n is to plot $\ln \varkappa$ versus $\ln M_n$

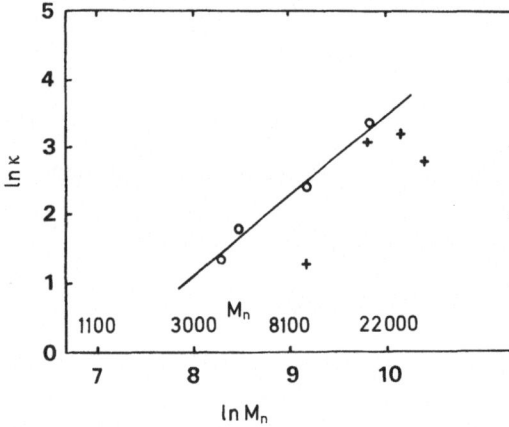

Fig. 19. Dependence of $\varkappa$ on M_n: tetrafunctional networks ($\bigcirc$) and trifunctional networks ($+$)

(ζ is assumed to be zero). This is done in Fig. 19. For the trifunctional networks the data are too scattered to permit a reliable conclusion. However, for the tetrafunctional networks the results can be well represented by

$$\varkappa = e^{-8.5}M_n^{1.2}$$

Thus $\varkappa$ doesn't seem to follow the predicted law. Nevertheless the scattering of data again encourages complementary measurements, particularly with swollen samples.

Acknowledgements: The authors are indebted to Professors P. J. Flory, B. E. Eichinger and B. Erman for their comments on this manuscript. Dr. J. P. Queslel also thanks Dr. C. W. Macosko for interesting suggestions about the branching theory and Dr. L. Garrido for stimulating discussions on rubber elasticity.

It is a pleasure to acknowledge the postdoctoral fellowship support generously provided J. P. Queslel by the Michelin Tire Company, and the financial support of this work by a grant from the National Science Foundation through its Polymer Program in the Division of Materials Research.

9 References

1. James, H. M., Guth, E.: J. Chem. Phys. *15*, 669 (1947)
2. Treloar, L. R. G.: "The Physics of Rubber Elasticity", 3rd ed., Clarendon Press, Oxford 1975
3. Flory, P. J.: J. Chem. Phys. *66*, 5720 (1977)
4. Flory, P. J.: "Contemporary Topics in Polymer Science", Vol. 2, Plenum Publishing Corporation, New York 1977

5. Rempp, P., Herz, J. E.: Angew. Makromol. Chem. *76/77*, 373 (1979)
6. Mark, J. E., Sullivan, J. L.: J. Chem. Phys. *66*, 1006 (1977)
7. Mark, J. E.: Pure Appl. Chem. *53*, 1495 (1981)
8. Mark, J. E.: Rubber Chem. Technol. *54*, 809 (1981)
9. Mark, J. E.: Adv. Polym. Sci. *44*, 1 (1982)
10. Langley, N. R.: Macromolecules *1*, 348 (1968)
11. Flory, P. J.: Proc. R. Soc. Lond. *A351*, 351 (1976)
12. Harari, F.: "Graph Theory", Addison-Wesley, Reading MA. 1971
13. Flory, P. J.: Macromolecules *15*, 99 (1982)
14. Scanlan, J.: J. Polym. Sci. *43*, 501 (1960)
15. Case, L. C.: J. Polym. Sci. *45*, 397 (1960)
16. Pearson, D. S., Graessley, W. W.: Macromolecules *11*, 528 (1978)
17. Graessley, W. W.: Macromolecules *8*, 186 (1975)
18. Flory, P. J.: Chem. Rev. *35*, 51 (1944)
19. Gottlieb, M., Macosko, C. W., Benjamin, G. S., Meyers, K. O., Merrill, E. W.: Macromolecules *14*, 1039 (1981)
20. Flory, P. J.: "Principles of Polymer Chemistry", Cornell University Press, Ithaca, N.Y. 1953
21. Flory, P. J., Rehner, J., Jr.: J. Chem. Phys. *11*, 521 (1943)
22. Flory, P. J.: J. Chem. Phys. *18*, 108 (1950)
23. Flory, P. J., Tatara, Y. I.: J. Polym. Sci., Polym. Phys. Ed. *13*, 683 (1975)
24. Mooney, M.: J. Appl. Phys. *11*, 582 (1940)
25. Mark, J. E.: Rubber Chem. Technol. *48*, 495 (1975)
26. Ronca, G., Allegra, G.: J. Chem. Phys. *63*, 4990 (1975)
27. Erman, B., Flory, P. J.: J. Chem. Phys. *68*, 5363 (1978)
28. Flory, P. J.: Polymer *20*, 1317 (1979)
29. Flory, P. J., Erman, B.: Macromolecules *15*, 800 (1982)
30. Erman, B., Flory, P. J.: J. Polym. Sci., Polym. Phys. Ed. *16*, 1115 (1978)
31. Pak, H., Flory, P. J.: J. Polym. Sci., Polym. Phys. Ed. *17*, 1845 (1979)
32. Erman, B., Wagner, W., Flory, P. J.: Macromolecules *13*, 1554 (1980)
33. Erman, B., Flory, P. J.: Macromolecules *15*, 806 (1982)
34. Flory, P. J.: Macromolecules *12*, 119 (1979)
35. Brotzman, R. W., Eichinger, B. E.: Macromolecules *15*, 531 (1982)
36. Mark, J. E.: Rubber Chem. Technol. *55*, 762 (1982)
37. Edwards, S. F., Brit. Polym. J. *9*, 140 (1977)
38. Gottlieb, M., Macosko, C. W., Lepsch, T. C.: J. Polym. Sci., Polym. Phys. Ed. *19*, 1603 (1981)
39. Bristow, G. M., Moore, C. G., Russell, R. M.: J. Polym. Sci. Part *A3*, 3893 (1965)
40. Moore, C. G., Watson, W. F.: J. Polym. Sci. *19*, 237 (1956)
41. Mullins, L.: J. Appl. Polym. Sci. *2*, 1 (1959)
42. Wood, L. A.: J. Research N.B.S. *80A*, 451 (1976)
43. Charlesby, A., Pinner, S. H.: Proc. Roy. Soc. Lond. *A249*, 367 (1959)
44. Langley, N. R., Polmanteer, K. E.: J. Polym. Sci., Polym. Phys. Ed. *12*, 1023 (1974)
45. Dossin, L. M., Graessley, W. W.: Macromolecules *12*, 123 (1979)
46. Bueche, A. M.: J. Polym. Sci. *19*, 297 (1956)
47. Falender, J. R., Yeh, G. S. Y., Mark, J. E.: J. Am. Chem. Soc. *101*, 7353 (1979)
48. Mark, J. E.: Makromol. Chem. Suppl. *2*, 87 (1979)
49. Sung, P. H., Mark, J. E.: Eur. Polym. J. *16*, 1223 (1980)
50. Sung, P. H., Mark, J. E.: J. Polym. Sci., Polym. Phys. Ed. *19*, 507 (1981)
51. Sung, P. H., Pan, S.-J., Mark, J. E., Chang, V. S. C., Lackey, J. E., Kennedy, J. P.: Polym. Bull. *9*, 375 (1983)
52. Mark, J. E., Rahalkar, R. R., Sullivan, J. L.: J. Chem. Phys. *70*, 1794 (1979)
53. Sharaf, M. A., Mark, J. E.: Rubber Chem. Technol. *53*, 982 (1980)
54. Macosko, C. W., Benjamin, G. S.: Pure and Appl. Chem. *53*, 1505 (1981)
55. Llorente, M. A., Mark, J. E.: Macromolecules *13*, 681 (1980)
56. Meyers, K. O., Bye, M. L., Merrill, E. W.: Macromolecules *13*, 1045 (1980)
57. Andrady, A. L., Llorente, M. A., Sharaf, M. A., Rahalkar, R. R., Mark, J. E., Sullivan, J. L., Yu, C. U., Falender, J. R.: J. Appl. Polym. Sci. *26*, 1829 (1981)

58. Opperman, W., Rehage, G.: "Elastomers and Rubber Elasticity", ed. by J. E. Mark and J. Lal, A.C.S., Washington, D.C. 1982, Chap. 16
59. Mark, J. E.: "Elastomers and Rubber Elasticity", Chap. 18
60. Kirk, K. A., Bidstrup, S. A., Merrill, E. W., Meyers, K. O.: Macromolecules *15*, 1123 (1982)
61. Llorente, M. A., Mark, J. E.: J. Chem. Phys. *71*, 682 (1979)
62. Smith, T. L.: "Elastomers and Rubber Elasticity", Chap. 22
63. Langley, N. R., Ferry, J. D.: Macromolecules *1*, 353 (1968)
64. Granick, S., Pedersen, S., Nelb, G. W., Ferry, J. D., Macosko, C. W.: J. Polym. Sci., Polym. Phys. Ed. *19*, 1745 (1981)
65. Valles, E. M., Macosko, C. W.: a) Macromolecules *12*, 673 (1979); b) Rubber Chem. Technol. *49*, 1232 (1976)
66. Eichinger, B. E.: two manuscripts submitted to J. Chem. Phys.
67. Llorente, M. A., Andrady, A. L., Mark, J. E.: J. Polym. Sci., Polym. Phys. Ed. *18*, 2263 (1980)
68. Meyers, K. O.: Ph. D. Thesis in Chem. Eng., M.I.T. (1980)
69. Leung, Y. K., Eichinger, B. E.: Preprints, Div. of Polym. Mat. *48*, 440 (1983)
70. Kennedy, J. P., Chang, V. S. C., Smith, R. A., Ivan, B.: Polym. Bull. *1*, 575 (1979)
71. Queslel, J. P., Mark, J. E., Fried, J. R., Lipsitt, B. M.: work in progress
72. Beshah, K., Mark, J. E., Ackerman, J. L.: Abstracts, Regional ACS Meeting, Miami University Ohio, May 1983
73. De Gennes, P. G.: J. Chem. Phys. *55*, 572 (1971)
74. Falender, J. R., Yeh, G. S. Y., Mark, J. E.: J. Chem. Phys. *70*, 5324 (1979)
75. Carpenter, R. L., Kramer, O., Ferry, J. D.: Macromolecules *10*, 117 (1977)
76. Ferry, J. D., Kan, H. C.: Rubber Chem. Technol. *51*, 731 (1978)
77. Pearson, D. S., Graessley, W. W.: Macromolecules *13*, 1001 (1980)
78. Ferry, J. D.: "Viscoelastic Properties of Polymers", 2nd Ed., Wiley, New York (1970)
79. Gee, G.: Macromolecules *13*, 705 (1980)
80. Gottlieb, M., Macosko, C. W.: Macromolecules *15*, 535 (1982)
81. Dossin, L., Pearson, D. S., Graessley, W. W.: Abstracts, International Rubber Conference, Kiev 1978
82. Macosko, C. W., Miller, D. R.: Macromolecules *9*, 199 (1976)
83. Miller, D. R., Macosko, C. W.: Macromolecules *9*, 206 (1976)
84. Miller, D. R., Valles, E. M., Macosko, C. W.: Polym. Eng. Sci. *19*, 272 (1979)
85. Bibbo, M. A., Valles, E. M.: Macromolecules *15*, 1293 (1982)
86. Graessley, W. W.: Adv. Polym. Sci. *16*, 1 (1974)
87. Graessley, W. W., Edwards, S. F.: Polymer *22*, 1329 (1981)
88. Carpenter, R. L., Kan, H.-C., Ferry, J. D.: Polym. Eng. Sci. *19*, 266 (1979)
89. Queslel, J. P.: Ph. D. Thesis, University of Paris VI, Paris 1982
90. Bye, M. L.: M. S. Thesis in Chem. Eng., M.I.T. 1980
91. Meyers, K. O., Merrill, E. W.: "Elastomers and Rubber Elasticity", Chap 17
92. Flory, P. J., Erman, B.: Macromolecules *16*, 595 (1983)
93. Eichinger, B. E.: Ann. Rev. Phys. Chem. *34*, 359
94. Queslel, J. P., Mark, J. E.: Preprints, Division of Polymer Chemistry, Inc., Am. Chem. Soc. *24* (2), 96 (1983)
95. Ullman, R.: Elastomers and Rubber Elasticity, Chap. 13
96. Ullman, R.: Macromolecules *15*, 1395 (1982)
97. Benoit, H., Decker, D., Duplessix, R., Picot, C., Rempp, P., Cotton, J. P., Farnoux, B., Jannink, G., Ober, R.: J. Polym. Sci., Polym. Phys. Ed. *14*, 2119 (1976)
98. Beltzung, M., Picot, C., Rempp, P., Herz, J.: Macromolecules *15*, 1594 (1982)
99. Beltzung, M., Herz, J., Picot, C.: Macromolecules *16*, 580 (1983)
100. Gottlieb, M.: J. Chem. Phys. *77*, 4783 (1982)

J. P. Kennedy (Editor)
Receivew July 18, 1983

Conformational Studies on Model Peptides

Their Contribution to Synthetic, Structural and Functional Innovations on Proteins

Franz Maser, Karsten Bode, V. N. Rajasekharan Pillai and Manfred Mutter
Institute of Organic Chemistry, University of Basel, CH-4056 Basel Swizerland

To improve our understanding of the vast variety of structural and functional features of peptides and proteins theoretical and experimental investigations of the interrelation between sequence, structure and function are necessary. This article reviews recent contributions to this field. First, the most popular methods for secondary structure prediction are discussed. In the light of the accuracy and limitations of these prediction algorithms *experimental* approaches for studying secondary structure formation are considered.

Systematic conformational studies of *homo*oligopeptides are shown to have provided valuable insight into the factors affecting secondary structure formation like chain length, side chain protecting groups, solvent, temperature and concentration. In addition, some recent results obtained from investigations of *co*oligopeptides are reported. In this context a newly designed host-guest approach appears to represent a particularly promising tool for the elucidation of the *sequence dependence* of secondary structure formation.

In the final section the usefulness of conformational predictions is pointed out. The reliable prediction of regular secondary structures formed in the early stages of protein folding is shown to be an important step towards the evaluation of the correct folding pathway. Moreover, as a consequence of the interrelation between sequence, conformation, and physico-chemical properties of peptides, conformational predictions are demonstrated to facilitate essentially the design of "tailor-made" peptides and the choice of the strategy to be applied in the synthesis of peptides. Finally, the use of conformational predictions supplementing computer-graphical methods in "rational drug design" is illustrated.

Advances in Polymer Science 65
© Springer-Verlag Berlin Heidelberg 1984

1 Introduction

A thorough understanding of the vast variety of reactions in which proteins participate in living cells requires information at a molecular level on the architectures of proteins of different origins and functions. Experimental and theoretical investigations on model peptides have been extremely useful in the interpretation of the three-dimensional structure of proteins and for the elucidation of the interaction between proteins and other systems. The fact that proteins can spontaneously regain a native state from the disordered state means that the rules to form a higher order structure enabling the function are derived from information about the primary sequence. Thus a systematic analysis to gain insight into the interrelation between the chemical detail of amino acid sequences and the conformational and functional features of the higher-order protein models should help the prediction of the three-dimensional structure of proteins from primary structure alone.

The tertiary structure of a protein results from a complex interplay of short-, medium- and longer range interactions occurring within the peptide chain. It is evident from the investigations on the helix-probability profiles of denatured proteins and their correlation with helical regions in native proteins that short-range interactions play the dominant role in determining the conformational preferences of the *backbones* of the various amino acids. On the other hand, the conformations of the *sidechains* of the native protein may be influenced by long- rather than by short-range interactions. Thus, the short-range interactions lead to one or more backbone nucleation sites along the peptide chain, where there is a tendency to form specific structures resembling the native conformations. These nucleation sites acquire additional stability from medium-range interactions and are directed toward each other to bring the long-range interactions required for the stabilization of the whole (backbone and side-chain) structure.

Experimental and theoretical approaches to the conformational studies on model peptides provide a basis for the understanding of the synthesis, structure and function of proteins on a molecular level. These studies aim to:

(i) elucidate the nature of the process that controls the folding of the polypeptide chains into the unique three-dimensional structures of proteins.

(ii) establish a basis for the design of polypeptides with specific structure and function.

(iii) understand the relationship between the amino acid sequence and the biologically active conformation.

(iv) find the correlation between physiochemical properties and the sequence; such a correlation can offer guidelines for an effective approach to the chemical synthesis and modification of peptide hormones and proteins.

(v) decode unknown protein structures using the empirical rules derived from the intercorrelation among the amino acid sequence, established structure and properties of known proteins.

The experimental approaches employed for these purposes involve the chemical synthesis and conformational studies of model peptide sequences (homo- and co-oligopeptides) and polypeptides, semisynthesis of proteins and the comparison of their function with natural proteins, and structure-activity correlation studies using peptide analogs. The conformational preferences of the model peptides in solution are usually determined by various spectral measurements. The x-ray structure analysis, which is the absolute method, can only be applied to cases where the substance can be obtained in crystalline state. In the case of proteins, this is unusual. The investigations in organic and aqueous solution are particularly important in view of the role of hydrophobic and hydrophilic interactions determining the native structure of proteins. The solubility of the oligopeptide sequences in suitable solvents, particularly in water, is often an obstacle to the investigations of the structure and properties of model peptides in solution. The present article is designed to review the major theoretical and experimental developments with regard to the important role of conformational properties of peptides and proteins. The authors do not intend to discuss every aspect thoroughly; instead, they prefer to select recent contributions to this subject.

2 Methods for Secondary Structure Prediction

2.1 Sequence and Secondary Structure

The secondary, tertiary and quartary structure of proteins determine their function, activity and specificity in the living cell. ANFINSEN demonstrated in a series of experiments that the process of protein folding occurs spontaneously without the need of any additional information beyond that coded in the amino acid sequence of the polypeptide chain [1,2]. Consequently, the primary structure defines the unique spatial arrangement, i.e. the so-called native conformation, of every residue. In principle, it should be possible to predict the protein structure from the amino acid sequence if this interrelation could be decoded.

Tertiary structure of proteins is considered to be the result of a complex interplay of short-range, medium-range and long-range interactions along the peptide chain. In a first approximation, the secondary structure is thought to be mainly influenced by short-range interactions. The evaluation of the relationship between primary and secondary structure therefore appears to be the first step on the way to unravel the interdependent parameters responsible for tertiary structure formation.

2.2 Ramachandran Map and Conformational Energy Diagrams

The backbone conformation of a peptide chain is characterized by the geometry of the amide bond $NH-CO$ and the dihedral angles $NH-C_\alpha H$ and $C_\alpha H-CO$ designated as ωi, Φi and Ψi for the ith residue, respectively [4]. The side chain conformation is described by $\varkappa i$ angles (Fig. 1).

Although the peptide unit is mostly considered to be plane ($\omega = 180°$) because of the partial double bond character of the amide bond, careful analyses by means of X-ray revealed deviations of torsion angles from planarity of up to 10°.

Despite the occasional irregularity of the ω-value, the most sensitive parameters to conformational changes are the Φ- and Ψ-angles. Ramachandran and coworkers demonstrated on the basis of "hard sphere models" that not every conceivable $\omega-\Psi$-combination is permitted. They proved that many arrangements result in steric hindrances within individual dipeptide units [5,6]. They constructed steric contour

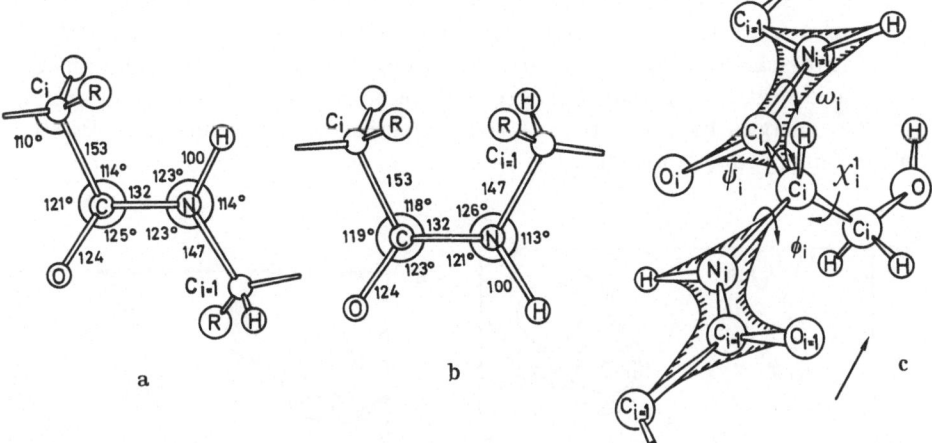

Fig. 1. a Geometry of the trans peptide bond [3], **b** Geometry of the cis peptide bond [3], **c** Representation of the dihedral angles [3]

diagrams ("Ramachandran maps"; Fig. 2) which graphically represent the allowed values of Ψ of a given residue i as a function of the corresponding Φ of every naturally occurring amino acid.

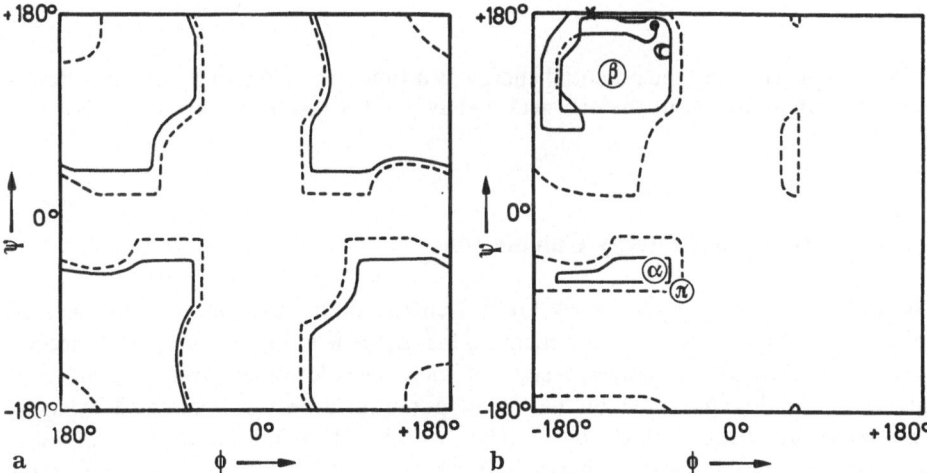

Fig. 2. Ramachandran diagrams **a** map for glycine, **b** map for alanine [3]

Regular secondary structures such as the different types of helices as well as parallel and antiparallel β-pleated-sheets, which can be considered helices with infinite rise per element, are characterized by a periodic succession of Φ/Ψ-values. These conformations are represented in the Ramachandran map by one point each corresponding to a well-defined Φ/Ψ-pair.

Irregular structures, however, give rise to a comparatively broad distribution of Φ/Ψ-angles spreading many points over the "allowed" regions of the map.

The hard-sphere model takes into account only steric aspects of each individual amino acid neglecting potential energy contributions. The inclusion of intraresidue potential energies by calculation of semi-empirical potential functions results in potential energy diagrams (Fig. 3).

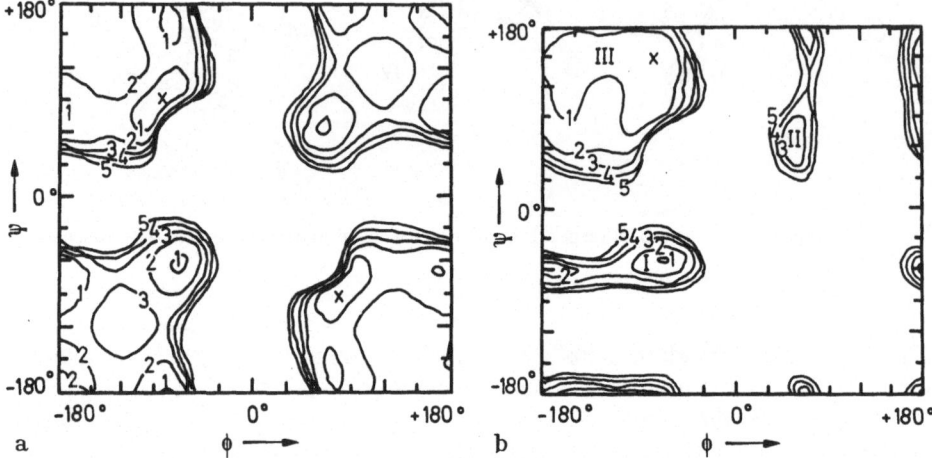

Fig. 3. Potential energy diagrams **a** map for glycine, **b** map for alanine [7]

These plots showing the potential energy as a function of Φ and Ψ provide quantitative informations about the energetic behavior of a residue within the "allowed" regions.

2.3 Conformational Energy Calculations

The conformation with the lowest GIBBS energy of the protein is established by calculations of the conformational energy. This energetic minimum is generally accepted to be related to the preferred state of the molecule under given external conditions[8]. The determination of the lowest potential energy is conducted by using semi-empirical functions which consist of contributions from non-bonded interactions, hydrogen bonding, intrinsical energy barriers of internal rotations and electrostatic interactions [6-13]. Since these terms are assumed to be independent they are treated separately. Moreover, all interatomic interactions are supposed to be additive. The computation of the energetic minimum is an approximation method because the complex solvent and hydrophobic effects have to be neglected, or they are treated in an simplistic way only.

Approximate quantum-mechanical attempts for the determination of the total energy of a molecule have also been reported [14]. However, these calculations were applied to small molecules only. Despite its flaws, the semi-empirical method basing on the laws of classical physics is of higher use for practical purposes. It can be extended to macromolecules more easily than the quantum-mechanical treatment.

For this reason, the computer program "ECEPP[1]" most frequently applied to the calculations of the conformational energy of peptides follows the semi-empirical approach.

Unfortunately, the evaluation of the energetic minimum of peptides is seriously complicated by the so-called "multiple-minima problem". This expression refers to the fact that the potential energy hypersurface generally features a large number of minima of comparable energy. Fig. 4 illustrates that different starting positions (I, II, III, IV) for the minimization procedure do not necessarily lead to the same minimum. The attempt of finding the "global" minimum therefore requires a reliable route to identify a "correct" and suitable starting conformation.

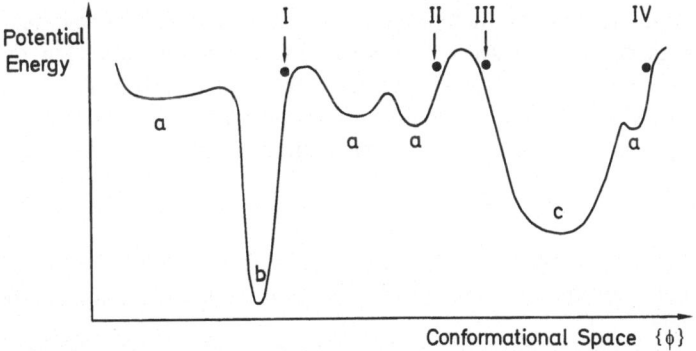

Fig. 4. Multiple-minima problem **a** local minima, **b** absolute minimum, **c** global minimum. The absolute minimum corresponds to a minimal energy; the global minimum corresponds to a minimal free energy (allowing for the entropy)

The multiple-minima problem has been successfully overcome for small oligopeptides and fibrous proteins [15]. Present efforts are aiming at the computation of the conformational energy of globular proteins by extending the semi-empirical methodology and by incorporating the effects of short-, medium- and long-range interactions [16].

2.4 Semi-Empirical Algorithms for Secondary Structure Prediction

Blout et al. were the first to report a defined relationship between amino acid sequence and secondary structure of polypeptides [17]. Referring to the results of conformational studies on synthetic homopolymers they designated residues with a significant preference for helical regions as "helixformers" whereas the others were called "helix-breakers".

---

1 *E*mpirical *C*onformational *E*nergy *P*rogramm for *P*eptides.

An increasing number of crystallographic data of proteins revealed that Blout's classification by and large held true for proteins, too. His observations along with ANFINSEN's hypothesis about a close interrelation between sequence and native conformation encouraged many groups to develop prediction algorithms for secondary structures. These methods can be subdivided into statistical, statistical-mechanical and stereochemical algorithms. However, this classification appears to be arbitrary since some approaches combine several routes differing from each other only with regard to points of emphasis.

2.4.1 Statistical Methods

The starting point of this technique is the statistical analysis of X-ray data of proteins. The known sequence and structure permit an assignment of each residue of the protein under consideration to a secondary structure element, e.g. α-helix, β-pleated sheet, and random coil (r.c.). The frequency of the amino acids in each of these categories is converted into a set of parameters for the probability of a given residue to occur in one of these three types of secondary structure. Deviating results in the literature are caused by differing criteria for the assignment of amino acids to the structural classes [18], by the delineation of the probability parameters and by the way these parameters are applied to the prediction of the conformation.

2.4.1.1 *CHOU-FASMAN Method* [19−23]

The algorithm developed and improved by Chou and Fasman is distinguished by its easily comprehensible procedure as well as by the fact that its simplest version can be used without a computer.

Chou and Fasman started out with the calculation of conformation preference parameters which represent a measure for the tendency of an amino acid to be part of an α-helical, extended or random coil region. Depending on the values of these parameters, they attribute each residue to one of the following six classes:

H_α = strong helix-former
h_α = helix-former
I_α = weak helix-former
i_α = indifferent to helix-formation
b_α = helix-breaker
B_α = strong helix-breaker

The authors set up a corresponding classification for β-structure formation (labeled with β-indices; Tab. 1). The increasing importance of β-turns for structural considerations resulted in methodically equivalent attempts to derive probability values P_t for amino acids to be involved in the formation of a bend.

The localization of secondary structures in proteins can be predicted by means of empirical rules making use of these conformation parameters. For example, a segment consisting of n residues (n ≥ 6) is predicted to adopt a helical conformation if $\langle P_\alpha \rangle \geqq 1{,}03$ and $\langle P_\alpha \rangle > \langle P_\beta \rangle^2$. However, additional conditions with regard to

---

2
$$\langle P_\alpha \rangle = \frac{\sum\limits_n P_\alpha}{n} \; ; \quad \langle P_\beta \rangle = \frac{\sum\limits_n P_\beta}{n}$$

Table 1. Conformational preference parameters P_α and P_β (based on data from 29 proteins [14])

Amino acid	P_α	Symbol	Amino acid	P_β	Symbol
Glu⁻	1.51	II_α	Val	1.70	H_β
Met	1.45	H_α	Ile	1.60	H_β
Ala	1.42	H_α	Tyr	1.47	H_β
Leu	1.21	H_α	Phe	1.38	h_β
Lys⁺	1.16	h_α	Trp	1.37	h_β
Phe	1.13	h_α	Leu	1.30	h_β
Gln	1.11	h_α	Cys	1.19	h_β
Trp	1.08	h_α	Thr	1.19	h_β
Ile	1.08	h_α	Gln	1.10	h_β
Val	1.06	h_α	Met	1.05	h_β
Asp⁻	1.01	I_α	Arg⁺	0.93	i_β
His⁺	1.00	I_α	Asn	0.89	i_β
Arg⁺	0.98	i_α	His⁺	0.87	i_β
Thr	0.83	i_α	Ala	0.83	i_β
Ser	0.77	i_α	Ser	0.75	b_β
Cys	0.70	i_α	Gly	0.75	b_β
Tyr	0.69	b_α	Lys⁺	0.74	b_β
Asn	0.67	b_α	Pro	0.55	B_β
Pro	0.57	B_α	Asp⁻	0.54	B_β
Gly	0.57	B_α	Glu⁻	0.37	B_β

helix-initiation, helix-propagation and helix-termination have to be fulfilled for the prediction of a segment to be helical [18]. The localization of β-sheets and β-turns can be achieved in a corresponding way.

A considerable step towards enhanced prediction accuracy was made by a more detailed analysis of the transition regions along the protein between different structural segments [19]. These experiments revealed that some amino acids exhibit pronounced positional preferences within one type of secondary structure. For example, L-Pro was found to occur twice as often at the N-terminus of an α-helix as an "average" residue ($P_{\alpha N} = 2.01$). On the other hand, it never occupies the C-terminal position of α-helices ($P_{\alpha c} = 0.00$).

Many efforts were made to refine the methodology of the prediction algorithms. The aid of computers became indispensable when known parameters were supplemented by values obtained from the statistical study of di- and tripeptide units [19,24]. Lifson and Sander discriminated between parallel and antiparallel β-pleated sheets [25], and Geisow and Roberts demonstrated that conformational preference parameters vary with the protein class [26]. Despite these refinements the improvement of the prediction accuracy must be considered minor. The upper limit of exclusively statistical algorithms appears to be in the order of 60–70% [27].

2.4.2 TANAKA-SCHERAGA Method [28–33]

The method by Tanaka and Scheraga will be exemplarily presented for the statistical-mechanical treatment of protein conformation because this approach has attained the highest popularity among all statistical-mechanical algorithms.

The first step involved is again the analysis of crystallographic data of a sufficiently large number of proteins. In the "three-state model" the conformational space of each residue is devided into n = 3 regions[3]. These three subdivisions correspond to α-helical (h), extended (ε) and coil (c) segments. The conformational properties of all 20 naturally occurring amino acids are described in terms of statistical weights which are expressed as relative weights referring to the c-state in order to reduce the number of parameters in the subsequent matrix treatment.

The statistical weight matrix constructed from the relative statistical weights permits the calculation of the partition function of a given sequence by matrix multiplication. The partition functions contribute to the evaluation of the conformational sequence probabilities $P(i/n/\{\varphi\}/)$, i.e. the probability to find a sequence of n residues in a specific conformational state $\{\varphi\}$ starting at the i-th position of the chain. The results of these calculations can graphically be presented in the form of conformational probability profiles.

Tanaka and Scheraga applied their "three-state model" to 19 different proteins with a prediction accuracy of 47–77%.

The peculiarity of this method is the potential to include *any* number of conformational states in the mathematical process. Cooperative effects can also be taken into account by calculations with a nearest-neighbor model.

2.4.3 Stereochemical Considerations

Approaches considering stereochemical aspects are the wheel representation of helical regions by Schiffer and Edmundson [34,45] and a secondary structure prediction scheme reported by Lim [36,37].

Schiffer and Edmundson observed that nonpolar residues within a helical region of a protein show a significant tendency to form nonpolar clusters. This finding was illustrated by means of a so-called "helical wheel" (Fig. 5).

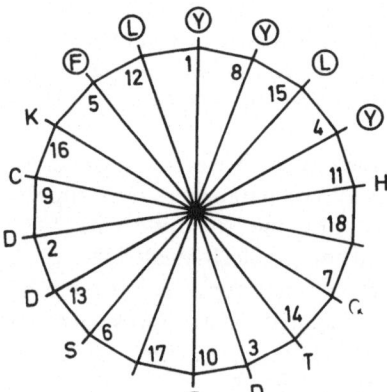

Fig. 5. Helical wheel representation of the C-terminus of adenylate kinase. Amino acid residues are characterized by one letter symbols. Nonpolar residues are encircled

---

3 The "three-state model" is a special case of a multi-state model in which n theoretically could be any number. A large number of conformational states involved in the prediction process generate a high degree of accuracy but it requires a considerable expenditure of computational time. It is for this practical reason that n is chosen to be 3 or 4.

For the prediction of helical regions it was suggested to check a given sequence for nonpolar triplets in the relative positions 1–4–5 and 1–2–5 which were considered potential nucleation centers for α-helices.

In a very elaborated procedure, Lim based the prediction of β-sheet structures on a discrimination of β-strands occurring on the surface and in the interior parts of a protein. Surface strands consist of alternating polar and apolar residues because they display a polar and nonpolar side. Internal strands, however, are built up by continuous stretches of nonpolar residues.

It must be emphasized that Lim's algorithm is considerably more sophisticated than outlined in the example above. His scheme comprise 22 and 14 rules for the prediction of α-helices and β-structures, respectively. On the other hand, this complexity must be regarded as its most serious shortcoming.

2.4.4 Algorithms Combining Different Aspects

The approaches briefly summarized in chapters 2.4.1–2.4.3 reflect the application of either various mathematical processes or different aspects of protein nature. But none of these methods turned out to be significantly superiour to another, and attempts to further improve them by combining their ideas are numerous.

Schulz [38] and Matthews [39] collected the results of several groups working in the field of secondary structure prediction and incorporated them into "joint prediction histograms". The prediction accuracy obtained by this procedure is comparable to the quality of the best individual prediction algorithm.

Palau and coworkers proposed a scheme with elements of purely statistical methods (conformational preference parameters) and structure-stabilizing factors ("weighting factors") [27]. The weighting factors modify the conformational preference parameters by taking into account e.g. hydrophobic interactions with β-structural regions or the occurrence of hydrophobic triplets in the helical positions 1–2–5 and 1–4–5. Additional parameters can be introduced into the prediction scheme.

Argos and Palau reported a marked correlation between the hydration potential of amino acids (as established by Wolfenden et al. [40]) and their positional preferences in β-structures [41]. This finding stresses the necessity to account for physico-chemical parameters of the residues involved in addition to probability and stereo-chemical factors.

2.4.5 Prediction Accuracy and Limitations

As mentioned in the previous chapter it must be seen that the results obtained from the various prediction schemes are quantitatively comparable [42]. It turned out that the success of a prediction algorithm depends on the protein class which it is applied to [2]. It also became evident that the prediction of α-helices can be carried out in a distinctly more accurate way than for β-pleated sheets or β-turns.

Despite the chance of localizing most helical regions and half of the β-strands and β-bends in a protein the major difficulty of precisely determining the termini of periodic secondary structures remains.

Admittedly, all these different approaches do not enable scientists to reliably predict secondary structure in any protein. The various routes which come to quantitatively almost identical results prove the presence of complex interactions which are not sufficiently understood yet.

Besides, it seems uncertain whether theoretical methods are applicable to biologically active peptides of small size and to model cooligopeptides under a variety of experimental conditions. These doubts stem from the facts that the derivation of algorithms bases on the analysis of native proteins disregarding solvent effects.

Consequently, there is a need of designing systematic experimental approaches for the elucidation of the laws governing secondary structure formation.

3 Experimental Approaches

3.1 Peptide Synthesis

Although eight decades have passed since Fischer started his pioneering work in the field of peptide synthesis [43,44], it still represents a formidable challenge to every organic chemist [45]. Predominantly five aspects illustrate the existing difficulties:

— The singular properties of each amino acid sequence prevent a reliable prediction of the physico-chemical behavior of the peptide from the starting materials [45].
— In general, only L-Ala and L-Leu among the 20 naturally occurring amino acids do not give rise to undersirable side reaction in the course of peptide synthesis [46].
— The most suitable combination of protecting groups has to be chosen individually for every sequence to be synthesized.
— The coupling method has to reflect kinetics and racemization processes during the synthesis [47]. The activation step as well as the appropriate choice of protecting groups is far from being routine.
— The strategy must be modeled after the goal of the synthesis. Products for medical purposes must be submitted to more drastic criteria than model peptides intended for conformational studies.

Progress with regard to methodology of peptide synthesis and purification stimulated the preparation of a large number of peptide hormones, protein partial sequences and model compounds for structural and functional investigations. The three most important strategies for the synthesis of medium-sized peptides are the

— classical method
— solid-phase peptide synthesis (SPS)
— liquid-phase peptide synthesis (LPS)

The classical approach [48,49] is characterized by the stepwise synthesis of short segments under homogeneous reaction conditions. The fragments are subsequently coupled via the segment condensation technique. The products are purified from side products and truncated sequences after each synthetic step. For this reason, products obtained via a classical method are distinguished by a high degree of purity and suitable for medical applications. On the other hand, this technique requires skilled chemists, and it is time-consuming. However, the gravest limitation of this approach is the generally low solubility of medium-sized peptides. These problems are of great weight as the chain-length of the peptide increases.

Merrifield designed the Solid-Phase Peptide Synthesis (SPS) [50-52] to overcome some of the shortcomings of the conventional procedure. The SPS was the first technique using a polymeric support. The carboxyl end of the growing peptide chain is covalently attached to an insoluble, derivatized polymer carrier. The peptide is extended towards the N-terminus in a stepwise manner by means of activation procedures familiar from the classical methods. The product is separated from reagents by filtration and obtained without intermediate isolation and purification. Consequently, quantitative yields in each coupling step are essential for a homogeneous final product. The application of excess activated reagents is to ensure a quantitative reaction.

The SPS overcomes solubility problems in the course of the synthesis, and it stands out for the simplicity and rapidity of the experimental steps as well as its potential to automatize the physical manipulations. Its major flaw is the accumulation of resin-bound impurities such as failure and truncated sequences because of incomplete coupling reactions [53-55]. In recent years, the SPS received new impulses by

— the development of new supports with superior swelling properties permitting an improved solvation of both matrix and growing peptide chain [56][57];
— the design of novel and more versatile anchoring groups ("multidetachable anchors") enhancing the flexibility of the synthetic strategy [58-60];
— progress in the field of chromatographic techniques such as preparative and semi-preparative HPLC [45].

3.2 Liquid-Phase Peptide Synthesis [61-65]

The Liquid-Phase Method (LPS) developed by Mutter and Bayer combines the advantages of a polymer-supported technique with those of a synthesis carried out under homogeneous reaction conditions. The C-terminal amino acid is coupled to a mono- or bifunctional poly(oxyethylene) (POE-M, POE), and the extension of the peptide towards the N-terminus takes place step by step. Unlike the solid-phase method, the LPS ensures coupling and deprotection in homogeneous solution. Excess low-molecular reagents are removed by precipitation of the POE-peptide with diethyl ether or by crystallization from ethanol or methanol.

The stability of the bond between peptide and its carrier can be varied according to the desired synthetic tactics. Usually, the terminal hydroxyl groups of POE are either converted into amino groups [66], or the support is derivatized by introducing appropriate anchoring groups [67-76] facilitating a mild cleavage of the product from the carrier.

POE features some physical properties which favor its application as an inert, macromolecular C-terminal protecting group in peptide synthesis:

— equivalence of the terminal functional groups [65];
— physico-chemical compatibility between peptide and support [72];
— high crystallinity [73-76];
— solubilizing influence [65,72,77];
— advantageous spectroscopic properties [65,72,76].

The last two aspects are worthwhile being outlined more explicitly.

Poly(oxyethylene) exerts a pronounced solubilizing effect on the peptides attached to it. Therefore, peptides are synthetically accessible which cannot be synthesized via the classical method because of solubility problems. Moreover, the conformation of POE-peptides can be studied at each stage of the synthesis in a large number of solvents covering a broad range of polarity. Due to the solubilizing influence of POE, CD-spectroscopic investigations of the conformation of hydrophobic homo-oligomers of L-Ala, L-Val and L-Ile were possible for the first time [78].

The favorable spectroscopic properties of POE are of great importance for conformational studies:

— POE is optically transparent up to the far-UV region [79], and it does not show a Cotton effect [76].
— The characteristic IR bands of POE do not interfere with the spectral regions important for peptide conformation (amide A, I, II, and V) [76].
— The ^{1}H- and ^{13}C-NMR signals of POE (resonances at 3.6 ppm and 71.5 ppm, respectively, against TMS) do not overlap with peptide signals relevant to conformational aspects [80–83].

Results obtained from conformational studies on POE-bound peptides and from investigations on analogous compounds under identical experimental conditions revealed that POE does not significantly influence the peptide conformation [76,84]. This property can be rationalized by the random coil conformation of POE in solution. Both low coil density and high segment flexibility hamper specific interactions between peptide and polymer [73].

Thus the liquid-phase method proved to be a powerful tool for the synthesis of medium-sized peptides as well as a helpful instrument for conformational studies employing a large number of spectroscopic techniques such as CD, ORD, UV, IR and NMR in a broad variety of solvents. These characteristics promoted the field of conformational investigations of oligopeptides [85].

3.3 Conformational Studies on Homooligopeptides

Conformational investigations on linear, monodisperse homooligopeptides proved to be a suitable route for a systematic examination of the individual parameters affecting the formation and stability of secondary structures [86–88]. Varying only one factor at a time the different contributions could be studied separately. Valuable information could be gathered about the role of chain length, solvent polarity, N-terminal protecting group, pH, ionic strength, temperature and concentration. Some of the results obtained from these studies will be presented in the following chapters.

3.3.1 CD Investigations

3.3.1.1 *Critical Chain Length and Role of the N-Terminal Protecting Group*

Figure 6 demonstrates the chain length dependence of the CD curves of Boc-(L-Met)$_n$-NH-POE (a) and $^-$Cl$^+$H$_3$N-(L-Met)$_n$-NH-POE (b) in 2,2,2-trifluoroethanol (TFE).

From Fig. 6A the subsequent conclusions can be drawn: up to a chain length of n = 6 the peptides adopt an essentially unordered conformation [89]. At the level of

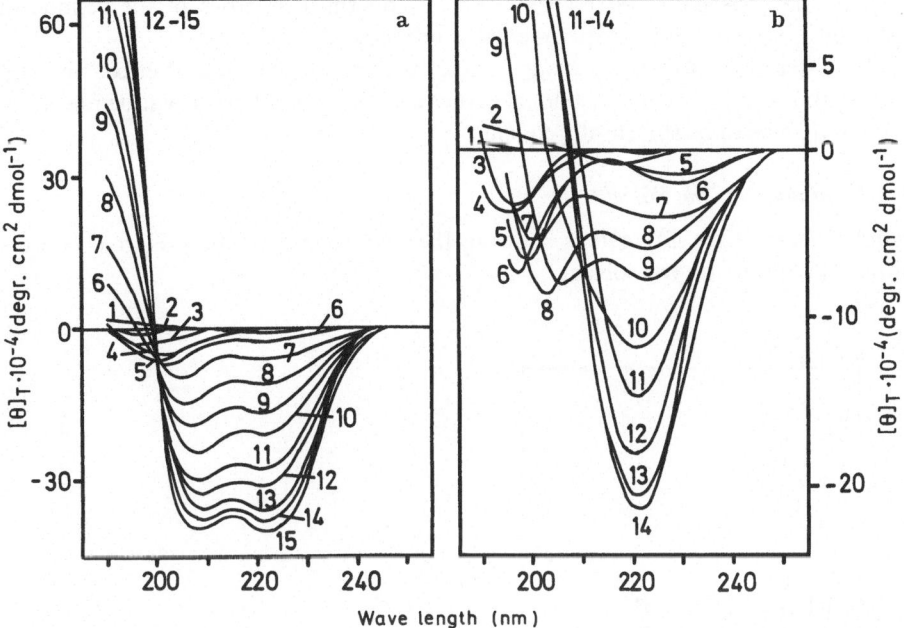

Fig. 6a and b. CD spectra of Boc (-L-Met-)$_{1-15}$NH-PEG **a** and $^-$Cl$^+$H$_3$N(L-Met-)$_{1-14}$NH-PEG **b** in TFE

the heptamer the onset of α-helix formation can be observed. The helix content grows as n increases [90].

The comparison of Fig. 6b with Fig. 6a impressively illustrates the drastic influence of the N-terminal Boc-protecting group on the preferred conformational state of (L-Met)$_n$ in TFE. The short oligomers (n = 1–6) of the N-protonated peptides exist in a predominantly unordered form. The CD curves of sequences with n = 7–9 are indicative of the coexistence of statistically coiled and ordered conformers [89]. The discrepancy becomes evident with higher oligomers (n = 10). The deblocked peptides adopt a β-sheet structure whereas, as mentioned above, the Boc-protected analogs form a partially helical conformation [89].

Similar conformational properties in TFE were established for the homooligomers of L-Glu (OBzl) [91] and L-Lys (Z) [92].

At a first glance, the results of CD studies on hydrophobic peptides of L-Ala, L-Val and L-Ile seem to be in conflict with those summarized above because both the Boc-protected and the N$_\alpha$-protonated derivatives of the longer oligomers form β-structures in TFE. In this solvent, the critical chain length for the random coil → β-structure transition was determined to be n = 7–8 [76,93,94].

However, a close look at the CD spectra reveals that the absence of the N$_\alpha$-Boc group stabilizes the β-sheet formation [95]. Therefore these findings parallel those reported for (L-Met)$_n$, [L-Glu(OBzl)]$_n$ and [L-Lys(Z)]$_n$.

The experiments described in this section illustrate:

— Homooligopeptides are apt to establish stable secondary structures after having exceeded a so-called "critical chain length" n$_c$. The type of conformation formed (particulatly α-helix- and β-sheet) depends on the properties of the residue of the oligomer and on the experimental conditions.

— The N-terminal protecting group generally exerts a pronounced effect on both type and stability of the secondary structure. The next chapter will demonstrate the influence to be solvent-dependent.

3.3.1.2 *Influence of the Solvent*

Figure 7 summarizes CD curves showing the dependence of the preferred conformation of peptides on the solvent.

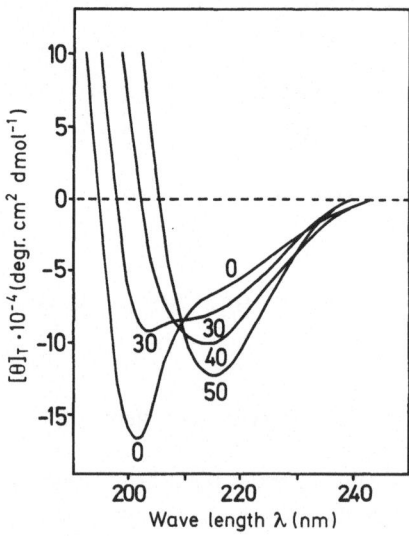

Fig. 7. CD spectra of Boc-(L-Ala)$_8$-NH-PEG-M in TFE and in solvent mixtures TFE/30% H$_2$O, TFE/40% H$_2$O, TFE/50% H$_2$O; c = 1,1 mg peptide/ml

The octapeptide Boc-(L-Ala)$_8$-NH-POE-M can be converted from a predominantly unordered state in TFE into an exclusive β-structure upon addition of water (TFE/H$_2$O (1:1 v/v)).

N$_\alpha$-protected homooligopeptides with a n$_c$-value of n$_c$ = 6–8 generally adopt a pleated sheet conformation in aqueous solution [34,45,45]. This result points to the importance of hydrophobic interactions for the stabilization of β-structures.

Methanol (MeOH) also has a tendency to support the formation of β-sheets. Studies on the homologous series of Boc-(L-Met)$_n$-NH-POE resulted in the adoption of an α-helix in TFE whereas in MeOH the β-structure was favored [34].

Systematic CD investigations of the influence of the solvent on peptide conformation lead to the following conclusions:

— Water and alcohols of comparatively low polarity such as methanol and ethanol are β-structure supporting solvents.

— Alcohols of intermediate polarity (2-chloroethanol, 3-chloropropanol- and 2,2,2-trifluoroethanol) are solvents suitable to stabilize the α-helical conformation; however, depending on the type of the peptide-constituent amino acid these solvents may as well tolerate β-structure formation.

— Alcohols of high polarity, e.g. 1,1,1,3,3,3-hexafluoroisopropanol (HFIP) tend to suppress β-sheets by competing with inter- and intra-peptide hydrogen bonds; these solvents are characterized by a pronounced hydrogen-bonding activity [89,94,95].

The sensitivity of the conformational properties of a given sequence towards polarity of the solvent system indicate that peptides may adopt different secondary structures in the nonpolar core or at the polar periphery of a globular protein. The observations outlined above might provide a plausible explanation for the problems encountered with the prediction of the preferred conformation of proteins unless information about the topological conditions are known in addition to the primary structure.

3.3.1.3 *Influence of pH*

Among others, Rinaudo and Domard were able to prove that the pH dramatically affects the conformational behavior of peptides. These authors studied oligomers of glutamic acid (n = 8–12) [90]. Sequences with completely neutralized side-chains were shown to adopt a β-structure whereas the charged analogs exist in a statistically coiled conformation. Between these two border-lines mixtures of disordered and ordered conformers were observed depending on the degree of neutralization. Higher oligomers at low pH were published to favor a helical structure, differing from the shorter analogs. Their properties in the fully charged form, however, remain unchanged.

Similar findings about the influence of the pH-value on peptide conformation were repeatedly reported [76,78,84].

3.3.1.4 *Role of Temperature and Ionic Strength*

A large number of CD studies confirm that temperature and ionic strength are two more factors important for secondary structure formation.

Contradictory observations with regard to the influence of temperature on the stability of β-structures [78,89,95] reflect two competing effects. The rise in temperature intensifies β-structure stabilizing hydrophobic interactions [96]; on the other hand, the more active Brownian motion causes the disruption of intermolecular hydrogen-bonding. The relative weight of the two contributions might be altered by the amino acids constituting the peptide under consideration.

An increase of ionic strength always acts in the direction of supporting β-structure formation by favoring hydrophobic interactions.

3.3.1.5 *Concentration Dependence of β-Structure Formation*

CD spectroscopy also proved to be useful for systematically studying the influence of the concentration on the conformation of oligopeptides [78,93,97].

Conformational transitions of the type β-structure → random coil can be compelled by diluting solutions of peptides with a potential to adopt a pleated sheet. This procedure permits the distinction between inter- and intramolecular β-structures ("cross-β-structures"). Cross-β-structures are characterized by a backfolding of the peptide chain; the intramolecular aggregation is insensitive to disruption by dilution.

The relative stabilities of β-structures were investigated by means of dilution experiments. Homooligopeptides of various residues were subjected to this approach suggesting the following order of increasing tendency of aggregation:

$$\text{L-Ala} < \text{L-Cys (Bzl)} < \text{L-Val} < \text{L-Ile}.$$

3.3.2 IR Investigations

Conformational analyses of peptides by IR spectroscopy[98] represent a valuable alternative to CD investigations. The latter are limited to solvents with an optical transparency in the absorption range of the peptide chromophor, i.e. with $\lambda \leq 250$ nm. This prerequisite makes available only a small number of solvents suitable for CD studies. Moreover, solvents relevant to peptide synthesis such as methylene chloride, chloroform, N,N-dimethylformamide and dimethylsulfoxide cannot be disposed of for CD spectroscopy. Additional problems with regard to the interpretation of CD spectra can be generated by overlapping curves resulting from aromatic side chains and aromatic protecting groups[99−108].

IR studies are generally not affected by the above-mentioned factors. The experiments described in chapter 3.3.1 can also be carried out by means of infrared spectroscopy. This method can optionally be applied in the solid state or in solvents which are optically transparent in the spectral regions of the conformationally sensitive amide bands (amide A, I, II and V).

Figure 8 illustrates the solid-state IR spectra of some interesting POE-bound $[\text{L-Lys-(Z)}]_n$ homooligomers. Short chains with $n = 1–3$ predominantly exist in a disordered conformation, oligomers with $n = 4–12$ adopt an antiparallel β-structure,

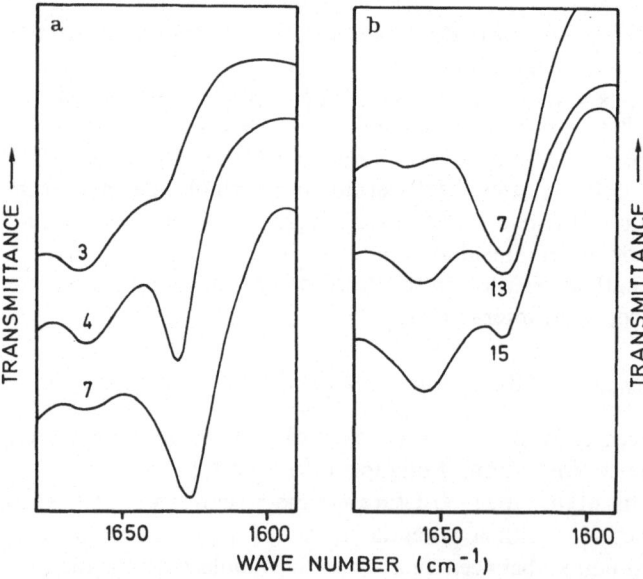

Fig. 8a and **b.** IR spectra in the 1680–1590 cm^{-1} region of PEG-bound- L-Lys(Z)$_n$ homopeptides in the solid state

whereas the spectra of the peptides with n = 13–15 are indicative of a coexistence of molecules in a β-sheet and α-helical conformation [92]. These results are in good agreement with observations made in the course of conformational studies on other homooligopeptides [104,105]. It can be concluded from these experiments that the critical chain length of β-structure formation in the solid state is n = 3–4; n_c for the onset of an α-helix under the same conditions appears to be n ≅ 13.

The particular importance of IR spectroscopy must be seen in its potential to discriminate between parallel and antiparallel β-sheets. CD measurements do not offer these advantages since only solid state experiments are possible for this purpose requiring considerable instrumental expenditures, e.g. the accessibility of the vacuum-UV region at λ ≥ 140 nm [106–110]. IR studies provide this insight under conventional operating conditions in both solution and the solid state [111–114].

Systematic investigations give evidence that homooligopeptides consisting of amino acids with linear side chains tend to adopt the antiparallel β-structure. Homooligomers with β-branched and aromatic residues, e.g. [L-Phe]$_n$ [107] and [L-Tyr(Bzl)]$_n$ [115] predominantly prefer the parallel β-sheet type [109]. [L-Leu]$_n$ with γ-branched amino acids was shown to adopt both of these structural variants [107].

3.3.3 NMR Investigations

The design of powerful NMR spectrometers and high-capacity computers as well as the development of advanced techniques stimulated the application of high-resolution NMR spectroscopy in the field of conformational analyses of peptides [10,116a]. ^{1}H-NMR spectroscopy, which this survey will be restricted to, is the most widely applied method but ^{13}C and ^{15}N are nuclei of growing interest [116b].

Nuclear magnetic resonance offers the advantage of allowing the observation of each individual residue in the peptide providing insight into the local conformation whereas optical techniques such as ORD, CD, UV and IR are confined to the study of overall molecular structures.

The most frequently used approach in ^{1}H-NMR experiments has been the investigation of the change of the chemical shift δ of the amide protons in dependence on temperature, concentration and solvent composition.

The determination of the temperature dependence of δ(NH) has been mostly carried out in solvents with strong hydrogen-bond acceptor properties, e.g. DMSO-d$_1$. Intramolecular hydrogen bonds result in a distinctly smaller temperature coefficient $\dfrac{d\delta}{dT}$ of the respective amide protons whereas hydrogen-bonding with the solvent is more easily affected by temperature changes giving rise to a larger $\dfrac{d\delta}{dT}$-value [80,83,116a–126]. As a general rule, a coefficient in DMSO-d$_6$ of $> 4 \cdot 10^{-3}$ ppm/°C is attributed to this latter kind of interaction; $\dfrac{d\delta}{dT}$ values of $< 2 \cdot 10^{-3}$ ppm are indicative of NH protons which are shielded from the solvent [116a]. — The study of δ(NH) as a function of peptide concentration supplies another approach to differentiate between intra- and intermolecularly hydrogen-bonded amide protons [125,127]. The disruption of interpeptide H-bonds by diluting a concentrated solution of the peptide under investigation results in an upfield shift of the NH-resonances which are affected by this procedure.

So-called titration experiments also proved to be sensitive to distinguish between amide protons exposed to and buried from the solvent [80, 83, 127−129]. A stepwise change of the polarity and protic character of the solvent system delineates the signals of NH's exposed to the solvent by a pronounced dependence of δ(NH) on the composition.

A number of more advanced methods were developed for conformational studies of peptides in addition to those listed above:

— determination of the vicinal coupling constant $^3J_{NH-C_\alpha H}$ which can be represented as a function of the dihedral angle Φ via a Bystrov plot [8, 116a, 130];

— measurement of the longitudinal relaxation time T, upon addition of stable radicals to the peptide solution. Exposed amide protons generate significantly broader signals because of their interaction with the radical [80, 131−133];

— saturation of exchangeable solvent resonances reduces the signal intensity of labile peptide protons via a process called "saturation transfer". Signals of

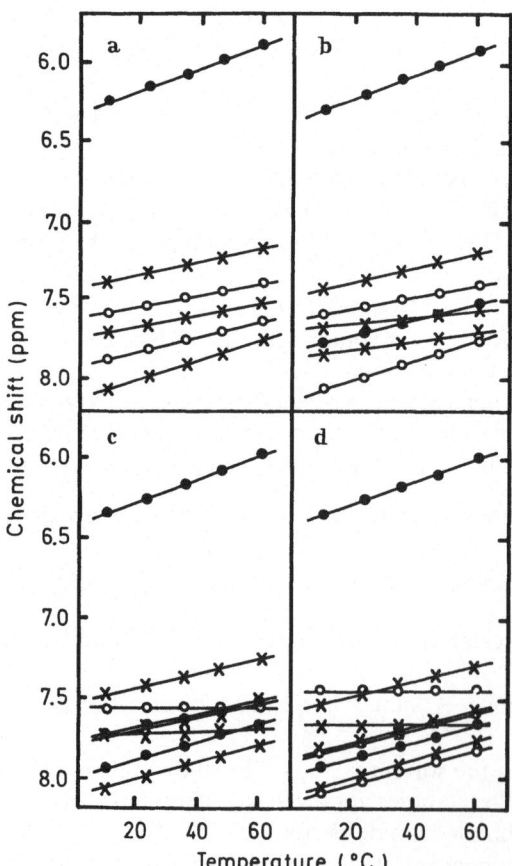

Fig. 9a—d. Temperature dependence of the NH chemical shifts of the PEG-bound methionine homooligopeptides Boc-(L-Met)$_n$NH-PEG-M (n = 5–8). Chemical shifts are referenced to internal TMS at all temperatures. **a** n = 5, **b** n = 6, **c** n = 7, and **d** n = 8. The upper curve corresponds to the urethane NH peak. The other curves are unassigned peptidic NH peaks (see text)

exposed, non-exchangeable protons grow more intensive as a consequence of the positive Nuclear Oberhauser Effect (NOE) [134-139];

— deuterium exchange experiments [116a,140,141].

Systematic investigations on homooligopeptides by means of ^{1}H-NMR spectroscopy were particularly stimulated by Goodman and coworkers [80,83,125,126,142-144]. Rapidly advancing progress in techniques suppressing very strong NMR resonances [80,83,142,145-147] permit the study of peptide conformation in protic, non-deuterated solvents as well as of sequences attached to POE. For this reason, the solubilizing effect of poly(oxyethylene) can be used for NMR experiments [80,83].

Goodman and Saltman subjected three series of POE-bound glutamate oligomers (n = 2–7) with different N-terminal protecting groups (Boc, Ac, pyroglutamyl) to extensive ^{1}H-NMR investigations. The authors demonstrated that, compatible with the results obtained by CD spectroscopy (chapter 3.3.1.1 and 3.3.1.2), the character of the N_α-protecting group and the polarity of the solvent exert a drastic influence on the conformation of the glutamates. On the other hand, a comparison of the ^{1}H-NMR spectra of Boc-(L-Met)$_n$-O-POE [148-150] and Boc-(L-Met)$_n$-OMe [83] does not indicate a comparatively influential role of the C-terminal group.

In Fig. 9, a plot of the chemical shift $\delta(NH)$ versus the temperature of the series Boc-(L-Met)$_n$-NH-POE (n = 5–8) in TFE is presented. The temperature coefficient $\frac{d\delta}{dT}$ of two amide resonances, particularly small for the hepta- and octamer, point to the presence of either buried NH's or amide protons involved in internal hydrogen bonding. This observation is in agreement with CD results. The latter gave evidence to the development of a partially helical conformation in TFE for n ≥ 7 (see Fig. 6). Considering the architecture of α-helices it becomes obvious that, for the heptamer, mainly two amide groups originating from the residues 4–7 are responsible for the stabilization of the helical structure [83]. This is an illustrative example for the superiority of ^{1}H-NMR spectroscopy when it comes to the evaluation of the *local* conformation.

3.4 Cooligopeptides

It has been shown in the previous section that systematic investigations on homo-oligopeptides enhance our understanding of the experimental parameters affecting secondary structure formation. Naturally, these studies must be unsuitable for evaluating the sequence dependence of peptide conformation. In order to extend the experiments to this important aspect a new concept has been designed. The host-guest approach, which will be discussed in detail in the subsequent chapter, replaces homooligopeptides by sequential peptides to systematically elucidate the interdependence between primary and secondary structure.

3.4.1 Host-Guest Technique

The terminology "host" and "guest" originated in the field of organic complex-chemistry [151]. Scheraga was the first to introduce it into peptide chemistry; he established the host-guest technique for the determination of helix-coil stability constants of all 20 naturally occurring α-amino acids [152-155]. The characteristic feature of his approach is the random incorporation of "guest" residues into an α-helical "host" *poly*peptide.

Goodman and coworkers applied a variant of this host-guest principle to the assignment of every N*H*- and C$_\alpha$H-signal in the high-resolution ^{1}H-NMR spectra of homooligo-(L-Met) by introducing single Gly guest residues into defined positions of the methionine host peptide [149].

Toniolo and Mutter et al. adopted the host-guest idea to study the impact to the guest amino acids glycine and L-proline on the preferred conformation of various host sequences [91,156–158].

These experiments emphasized that L-Pro acts as a "helix-breaker" when inserted into the center part of a peptide of α-helical structure. However, placed at the N-terminus, proline behaves like a "helix initiator". These findings, in perfect agreement with results obtained from statistical analyses of proteins (see chapter 2.4.1.1), can be rationalized by stereochemical studies on dipeptide units of the type L-Pro-L-X and L-X-L-Pro [91].

With regard to β-structures, L-Pro was found to exert a destabilizing influence due to the introduction of structural irregularities. Glycine as a guest residue was shown to hamper the formation of periodic secondary structures such as helices and β-sheets because of its peculiar conformational flexibility.

Both Pro and Gly have to be considered unusual among the 20 naturally occurring α-amino acids. Their significant influence on the conformation of a host peptide could a priori be expected from a look at the conformational energy maps (Ramachandran maps) of these two residues. To make it possible to extend the host-guest principle to amino acids with less dramatic conformational behavior, the original approach was modified such as to insert two vicinal guest residues into the host sequence.

Figure 10 explains the basic idea behind this.approach.

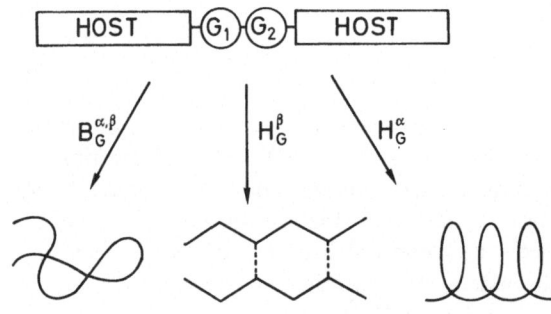

Fig. 10. Principle of the host-guest technique

Two guest amino acids (G) are incorporated into two adjacent positions of a host peptide which is potentially able to exist in a random coil, α-helical and β-sheet conformation, depending on the experimental conditions. The guest amino acids are supposed to alter the conformational state of the sequence reflecting the structural preferences of the "guest":

— Host-guest peptides with guest residues having a pronounced preference for the α-helix (designated by "H_G^α") are expected to adopt a helical conformation under helix-promoting conditions.

— Correspondingly, host-guest peptides containing guest residues with a high β-sheet potential ("H_G^β") are to form a β-structure.

— Guest amino acids with a high tendency to disrupt ordered structures ("B_G^β") should cause a conformational transition to the random coil state.

Various reasons advised the choice of (L-Ala)$_n$ as the host peptide:

— Conformational energy diagrams of L-Ala show the presence of minima for Φ/Ψ-dihedral angle combinations representing both α-helix and β-structure. The energy of these two minima is about the same.

— Intensive conformational investigations proved that (L-Ala)$_n$ can adopt a random coil conformation, a β-sheet structure or a partial helix depending on the experimental conditions applied as well as the chain-length n and the nature of the protecting groups [78,79,91,95,114,159−163].

— Conformational energy calculations resulted in a large number of energetically comparable minima for (L-Ala)$_{10}$ [164,165].

These findings suggested homooligo-alanine to be of high conformational versatility. This is an important prerequisite for its usefulness as a host peptide which should ideally be able to reflect the conformational preferences of the guest residues.

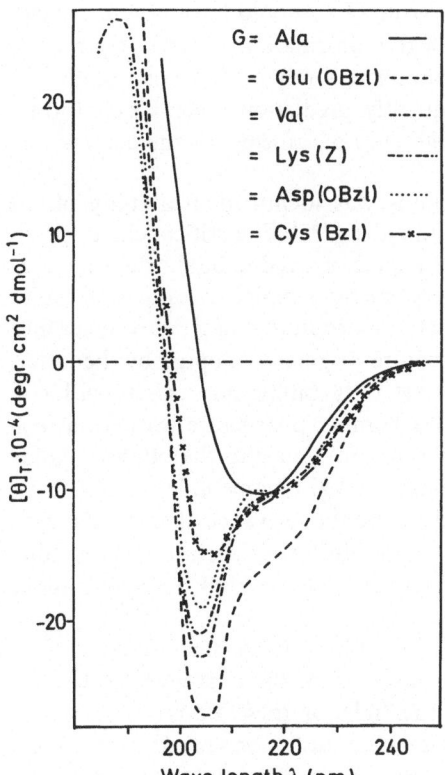

Fig. 11. CD spectra of the host-guest peptides Boc-L-Ala$_5$-G$_2$-L-Ala$_3$-NH-PEG-M in TFE, c = 1,1 mg peptide/ml

A look at Fig. 11 reveals the drastic influence of the guest residues on the conformation of the host peptide in TFE. Whereas (L-Ala)$_{10}$ adopts a pure β-structure, insertion of two guest amino acids either entirely or partially disrupts the ordered structure. Guest residues with a distinct helix-inducing potential replace the β-structure by a partially α-helical conformation. This observation holds true for L-Glu (OBzl), L-Val, L-Lys (Z) and L-Asp (OBzl) (Fig. 11). However, the L-Cys (Bzl)-containing host-guest peptide remains in a partial β-structure as can be deduced from the CD spectrum [166].

Interpreting the CD spectra with regard to the α-helix content of the host-guest sequence in a semi-quantitative way, the following tentative scale for the α-helix potential in TFE of the guest residues examined could be derived:

$$\text{L-Glu (OBzl)} \gg \text{L-Val} > \text{L-Lys (Z)} > \text{L-Asp (OBzl)} \gg \text{L-Cys (Bzl)}$$

Neglecting minor deviations, this order is in harmony with scales established on the base of statistical conformational parameters of the corresponding side chain-unprotected residues [19].

At a first glance, this result seems to generally support the validity of the conformational preference parameters in prediction schemes. However, the statistical analysis of proteins favors the β-structure potential of L-Val over its helix-inducing power. Provided these theoretical prediction methods could be applied not only to proteins but also, in a first approximation, to synthetic oligopeptides, a stable β-structure for Boc-(L-Ala)$_5$-(L-Val)$_2$-(L-Ala)$_3$-NH-POE-M in TFE should have been expected. The experimental outcome of a partial α-helical conformation for this sequence in TFE points to limitations of the prediction rules which rely on the assumption of a dominance of short-range interactions. Consequently, prediction of peptide conformation requires more informations than the preference parameters of the constituting amino acids alone.

The importance of medium-range interactions was identified in a study of the β-sheet forming potential of various guest amino acids [167]. The critical chain length of β-structure formation of the host-homooligopeptide proved to be shorter than that of any sequential host-guest sequence in every medium under investigation (solid state, CH$_2$Cl$_2$, MeOH, H$_2$O, TFE). Evidently, the incorporation of guest residues into a host peptide generally destabilizes the β-structure originally adopted by the latter. More detailed investigations indicated that not only bulky side chains within a sequence of spatially unpretentious side chains hamper β-structure formation (e.g. L-Val, guest residues in a (L-Ala)$_n$ host sequence), but that also the opposite holds true (L-Ala guest amino acids incorporated into a L-Val oligopeptide).

With the experiments it was possible to underscore the relevance of stereochemical criteria, i.e. medium-range interactions, for the stability of β-structures. The prediction of β-sheets on the exclusive basis of algorithms of the kind described above must necessarily be of inferior quality.

Delineation of conformational preferences of individual residues under various conditions, which can be considered the original goal of the host-guest technique, seems to be of subordinate importance in the light of these findings. However, this approach in addition opens the door to systematic studies on sequence-dependent parameters and stereochemical factors influencing secondary structure formation.

The following schemes outline some of the experimental opportunities of this model.

— Importance of the position which the guest residue (*G*) occupies in the host peptide. From these investigations asymmetric conformation nucleation properties of the guest amino acid might possibly be derived.

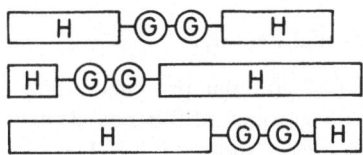

— Host guest peptides with three-functional guest residues can be used for studies of the effect of various side chain protecting groups and combinations of blocking groups on conformational parameters. As a consequence of a close relationship between solubility and preferred peptide conformation, experiments designed for this purpose can be of utmost relevance with regard to the strategy of peptide synthesis (see chapter 4.2).

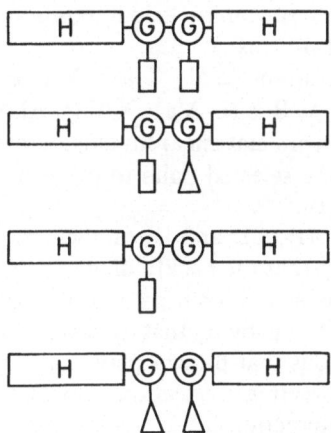

— Relative β-turn formation potentials of selected dipeptides can be determined and compared with corresponding values of prediction codes.

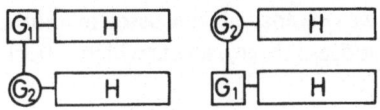

— Conformational consequences of bulky guest residues for the host peptides can be examined under α-helix- and β-structure-promoting conditions.

— Interesting results may also be obtained by varying the number of guest residues incorporated into the host peptide.

These schemes illustrate the versatility of the host-guest technique. It must be considered an approach of great promise for the elucidation of the sequence-depending rules governing peptide and protein conformation by singling out individual contributions of factors which affect secondary structure formation in a complex way.

3.4.2 Investigation of β-Turn Forming Model Peptides

β-turns are important structural elements of globular proteins. About one third of all amino acids constituting this protein class is part of β-bends. β-turns are built up by four consecutive residues designated with the indices i, i + 1, i + 2, i + 3[4]. The geometry of this structure is characterized by a reversal of the peptide chain by about 180°. Venkatachalam [168] and Lewis [169] classified the β-bends into 6 and 11 different types, respectively, according to the Φ/Ψ-dihedral angles of the two amino acids occupying the corner positions i + 1 and i + 2. In most cases the chain reversal is stabilized by a hydrogen bond between NH^{i+3} and CO^i (so-called "4 → 1 bridge") [169], but ionic [170] and covalent stabilizations are also reported [171,172].

Brahmachari et al. examined tripeptides of the general formula Ac^1-L-Pro^2-Gly^3-X^4-OMe by altering the amino acid X in position i + 3 (X = Gly, L-Ala, L-Leu, L-Ile, L-Phe) [173]. Their studies demonstrated that in addition to the well-established influence of the residues i + 1 and i + 2 also the amino acid in position i + 3 exerts a noticeable contribution to the stability of β-turns.

Bode et al. incorporated pairs of guest amino acids X and Y into an alaninic host peptide resulting in a sequence of the type Boc-(L-Ala)$_2$-X-Y-(L-Ala)$_2$-NH-POE. The choice of X and Y referred to results from statistical investigations on proteins which indicated different probabilities of the selected pairs to occur in the relevant positions i + 1 (X) and i + 2 (Y) of a β-turn [174].

Spectroscopic studies by means of ^{1}H-NMR, CD and IR on these model peptides confirmed the prediction codes [175] implying that the backfolding of peptide chains is chiefly determined by short-range interactions. This conclusion is in good agreement with results from similar investigations [176,177] proving that even very short peptides are able to form a chain reversal. It is obvious that β-turns play an active role in the folding process of proteins possibly by directing nucleation centers towards each other under the influence of short-range interactions.

4 Scope of Conformational Predictions

Conformational prediction is a fascinating subject, and it is related to a vast area of application. The usefulness of this field was extensively discussed in another review [23], and the subsequent three chapters will address three aspects which reflect the authors' particular interest.

---

4 The positions are designated starting from the N-terminus. An alternative labeling system is by Figs. 1, 2, 3, 4.

4.1 Secondary Structure and Folding Mechanism [179a)]

The renaturation process of globular proteins takes place within 10^{-1} and 10^{-3} seconds. Internal molecular rotations are known to be in the order of 10^{-13} seconds [179b)]. Taking into account only three energetically favorable conformational states for each amino acid, a relatively small protein of 150 residues can potentially adopt more than 10^{60} conformations. A comparison of these figures illustrates in an impressive manner that the time available to a protein for the folding process is by far not sufficient to find the energetically most favorable conformation by a random search mechanism.

The native conformation is assumed to be the result of a structural self-organization process which starts in various regions of the chain independently and simultaneously. The idea behind this mechanism is the initiation of protein folding by developing local, fluctuating secondary structures [180−184)] as a consequence of short-range interactions [185−187)]. The stabilization of the "nucleation centers" formed in the initial stage is achieved by interactions arising in the course of the subsequent step of self-organization. In this context, a propagation mechanism [179b,184,188,189)] and/or diffusion-and-collision mechanism [183,190,191)] are discussed. Influenced by long-range interactions the existing structural units assemble giving rise to the formation of independent domains [189,192−194)], in which the polypeptide chain by and large adopts the native conformation. A final assembly of these domains results in the native tertiary structure [179a)].

The evaluation of the correct folding pathway is expected to be one possible way to overcome the multiple-minima problem encountered with conformational predictions on the basis of energy calculations [42)]. A long first step towards this goal is the reliable prediction of the regular secondary structure formed in the early stages of protein self-organization. For this reason, experiments having in view the understanding of the mechanism of secondary structure formation also represent valuable contributions to the prediction of the native conformation of proteins from the amino acid sequence.

4.2 Relationship Between Sequence, Conformation and Physico-Chemical Properties of Peptides

The present state of knowledge about the relationship between primary structure and physico-chemical properties of peptides does not allow an immediate prediction of physico-chemical parameters from the peptide sequence:

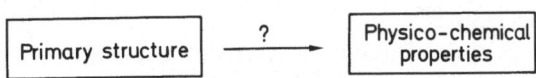

On the other hand, conformational studies on homo- and cooligopeptides revealed an intimate correlation between the preferred conformation of a peptide and its physico-chemical behavior [156,158,195)]. Conformational investigations in the course of peptide synthesis disclosed that frequently transitions from a "statistically" coiled conformation to ordered structures occur as a result of further chain elongation. These transitions manifest themselves in drastic changes of the viscosity of the reaction medium, the solubility of the peptide and the reactivity of the terminal amino groups. Fig. 12 summarizes semi-quantitative observations concerning the chain length dependence of the degree of solubility of oligopeptides.

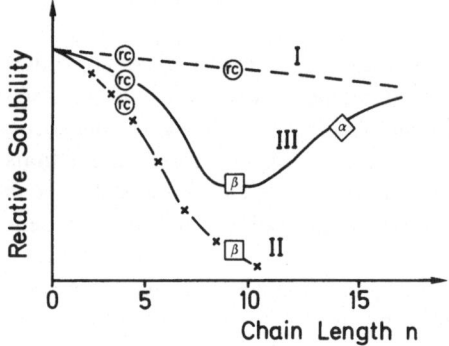

Fig. 12. Solubility and conformation of oligo-peptides

Homooligopeptides of hydrophobic amino acids such as L-Ala, L-Val and L-Ile undergo a random coil → β-structure-type transition at chain lengths of $n = 6$–8. This transition is paralleled by a significant drop of the solubility (curve II) and a reduced reactivity of the terminal amino groups. Therefore the stepwise synthesis of peptides is occasionally restricted to chain lengths of $n = 8$–10.

In the course of the synthesis of homoologopeptides with α-helix-forming residues, e.g. L-Glu (OBzl), L-Lys (Z) and L-Met, the solubility curve goes through a minimum at $n = 7$–11. Upon further chain elongation, however, an increase of the solubility is observed (curve III). This behavior can be explained by a r.c. → β-type transition at $n \cong 7$ with a subsequent β-structure → γ-helix transition at $n > 11$.

The solubility properties of oligopeptides which do not adopt any secondary structure during the synthetic process are schematically represented by curve I.

Apart from the observations discussed above, there is additional evidence for a close relationship between solubility and conformation of peptides:

— IR studies revealed that N_α-protonated oligopeptides have a significantly more pronounced tendency to adopt the β-structure in CH_2Cl_2 than their N_α-Boc protected analogs. This finding is in harmony with experiences in peptide synthesis: Once the N_α-Boc groups are removed the solubility of the resulting protonated peptide decreases drastically [196].

— It is conventional in peptide synthesis to start out in a solvent of low polarity (e.g. CH_2Cl_2) and to add solvents of higher polarity such as DMF and DMSO as soon as the solubility of the growing peptide chain falls off. The addition of DMSO or DMF causes a disruption of the β-sheet structure as monitored by the lowered intensity of the amide-I IR band at 1630 cm^{-1} (Fig. 13) because this band is indicative of the presence of an antiparallel β-structure.

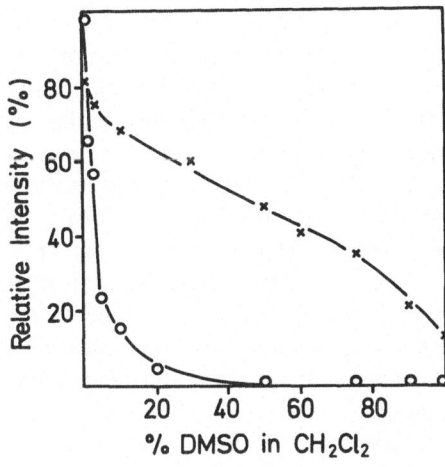

Fig. 13. Relative intensities (% area) of the 1630 cm⁻¹ amide I-band of Boc-(L-Ala)$_6$-NH-PEG-M (—x—) and Boc-L-Ala-L-Val$_2$-L-Ala$_3$-NH-PEG-M (—o—), depending on solvent composition

The correlations between physico-chemical properties and peptide conformation described in this chapter give in principle access to an indirect route for the prediction of the physico-chemical behavior of peptides from their primary structure:

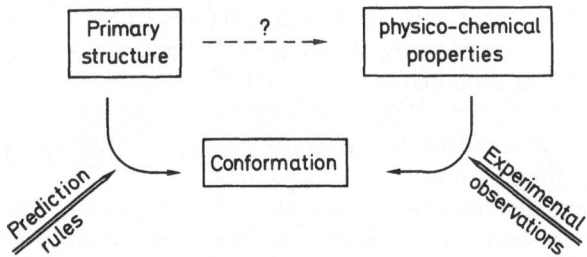

This approach, however, requires the elaboration of a reliable algorithm for the prediction of peptide conformation from the sequence. Peptides with "tailor-made" conformational and physico-chemical properties could be the consequence.

A profound knowledge of the interrelation of sequence, conformation and physico-chemical properties would be of high interest with regard to the choice of the optimal strategy for the synthesis of a given peptide. Bearing in mind conformational aspects the sequence should be divided into several medium-sized fragments, which subsequently had to be condensed to the product with increased solubility (see Fig. 12, curve III). Therefore, reflections giving due consideration to the conformational as well as the physico-chemical properties of peptides should help overcome solubility problems, which are one of the major obstacles in peptide synthesis.

4.3 Conformation and Biological Activity

Years ago, the potential of conformational predictions for the elucidation of the biological activity of polypeptides has been recognized [19,178]. Occasionally, the direction of conformational changes of a protein in response to the variation of experimental conditions could be confirmed by prediction methods. The regions involved in conformational activities were shown to frequently have high tendencies to potentially adopt various types of secondary structures, giving preference to the one conformation which is stabilized by the respective environment.

Maybe the most ambitious aspect of current conformational studies on peptides is the correlation to their biological activity, which is determined by the conformation of the protein when it is interacting with the receptor ("receptor-bound conformation"). Systematic introduction of chemical modifications with *predictable* conformational effects in the strategy employed helped to identify the receptor-bound conformation. Retention of biological activity implies that the conformational constraints imposed by the alteration are tolerated by the receptor. Research in this field can be summarized by the expression "rational drug design".

An illustrative example for studies of the interrelation of the primary structure, conformation and biological activity was reported recently [197]. The authors studied the α-mating factor, a tridecamer with sequence Trp^1-His^2-Trp^3-Leu^4-Glu^5-Leu^6-Lys^7-Pro^8-Gly^9-Glu^{10}-Pro^{11}-Met^{12}-Tyr^{13} which specifically inhibits DNA replication in cells [198,199]. Both the native peptide and a number of analogs were synthesized with regard to their conformation-activity relationship. No reduction of activity was observed in the absence of the N-terminal Trp^1 but the deletion of the two residues Trp^1 and His^2 rendered the peptide inactive. On the other hand, the successive shortening of the C-terminus gradually lowered the activity. The truncated nonamer Trp^1-Gly^9 was inactive. Consequently, the shortest unit essential for the retention of the biological activity was proposed to be His^2-Glu^{10}.

The physiological properties of the α-mating factor and its analogs could be explained by means of conformational considerations. ^{1}H-NMR investigations revealed a folded conformation of the native peptide.

The structure is characterized by the presence of three β-turns. The N-terminal Trp^1 is not required for retaining the folded conformation but when both Trp^1 and His^2 are removed the N-terminal chain reversal is destroyed causing the loss of activity.

This example illustrates the usefulness of conformational studies for the elucidation of biological activity but it also justifies the desire to establish a reliable correlation between primary and secondary structure allowing chemical modifications with predictable conformational effects.

Another up-coming field in conformation-activity studies is the application of computer graphical methods [200,201] such as the Merck Molecular Modeling System (MMMS) [200,202]. These techniques turned out to be a valuable tool for the design of conformational constrained analogs of biologically active molecules. These compounds, e.g. cyclic peptides [116,203,204], possess an artifically introduced restriction of their conformational flexibility by modifying the peptide backbone. This approach allows a more detailed investigation of the influence of the conformation on biological activity.

A particularly advantageous feature of computer graphical methods is the chance of three-dimensional superposition of two molecules for structural comparison. For example, Freidinger et al. identified a close conformational similarity between type-II^1 β-turns and segments in which the residues i + 1 and i + 2 were replaced by γ-lactames [205]. This finding was applied to a LH-RH analog. The previously postulated type II^1 β-turn of this hormone was substituted by a conformation-restraining γ-lactame group enhancing the bioactivity of the modified compound both in vivo and in vitro [201,205].

Another impressive example for the advantages of computer graphics in synthesis planning was the design and subsequent synthesis of a highly active hexapeptide analog of somatostatin by Veber [206]. Based on some earlier work, a model of the active conformation of somatostatin was postulated [207]. In addition, a bioactive bicyclic analog was conformationally defined by means of MMMS. These experiments culminated in the hypothesis that the β-turn region of this compound probably contained all information required for physiological activity. Using MMMS again, one of the two cyclic units was replaced by various bridging elements. Different types of β-turns were found to be suitable for this purpose. Therefore the tetrapeptide sequence Phe-D-Trp-Lys-Thr was supplemented by systematically varied dipeptide bridges with a high β-turn potential according to both sterical and statistical aspects. Among the resulting cyclic hexapeptides, cyclo-(Phe-D-Trp-Lys-Thr-Phe-Pro) turned out to be of high biological activity. This relatively easily accessible compound proved to be even active upon oral application.

5 Outlook

The article presented here indicates the fast growing number of contributions to the area of conformational properties of peptides. Despite considerable progress in the understanding of limited aspects there is still quite a way to go. It became evident in this review that theoretical prediction and experimental results are occasionally incompatible, and more data have to be collected. But molecular modeling and computer-assisted energy calculation prove to be useful when it comes to designing model experiments.

The urgent desire of chemists to predict the conformation of a protein from its primary structure will exist for some more time. On the other hand it is now conceivable that any type of secondary structure can artificially be constructed. This accomplishment might be of interest for defined replacements of ordered regions in proteins by synthetically easily accessible oligomers of the same conformation. Most notably, the understanding of secondary structure formation is a first step on the way to synthesize more complex polypeptide structures; as a final goal, the design of folded polypeptide domains with specific physicochemical and biological properties (Fig. 14) will open up new perspectives not only for the peptide chemist, but, on a long run, even in the field of genetic engineering.

Fortunately, both the creativity of chemists and the improvements with regard to instrumentation and synthesis of peptides will guarantee a steady flow of informations which are necessary to enlarge and, possibly some day, complete the picture of the rational standing behind the conformational characteristics of peptides and proteins.

Perspectives for Peptide Synthesis

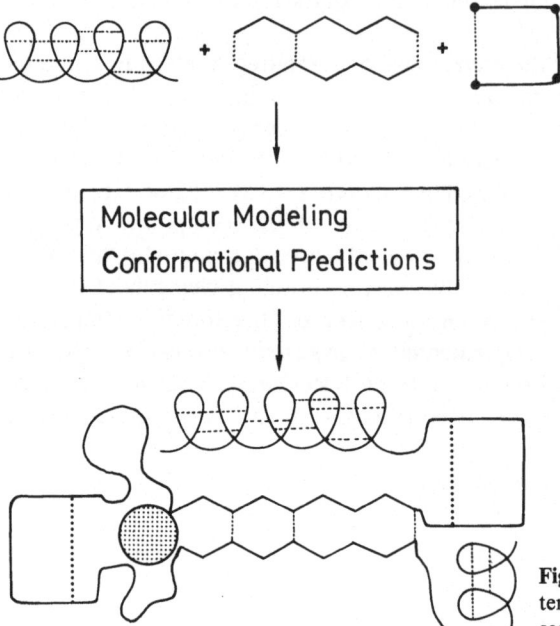

Fig. 14. Assembly of a polypeptide with tertiary structure from segments with secondary structure

Acknowledgement. This work was supported by the "Deutsche Forschungsgemein-schaft" (SFB41).

6 References

1. Anfinsen, C. B., Haber, E., Sela, M., White, Jr., H. F.: Proc. Natl. Acad. Sci. USA *47*, 1309 (1961)
2. Anfinsen, C. B., Scheraga, H. A.: Adv. Prot. Chem. *29*, 205 (1975)
3. Schulz, G. E., Schirmer, R. H.: Principles of Protein Structure Springer Verlag, New York, 1979, S. 86
4. IUPAC-IUB Commission on Biochemical Nomenclature 1969, Biochemistry *9*, 3471 (1970)
5. Ramachandran, G. N., Ramakrishnan, C., Sasisekharan, V.: J. Mol. Biol. *7*, 95 (1963)
6. Ramachandran, G. N., Sasisekharan, V.: Adv. Protein Chem. *23*, 283 (1968)
7. Flory, P. J.: Statistical Mechanics of Chain Molecules, Interscience, New York, 1969
8. Némethy, G.: Biochimie *57*, 471 (1975)
9. Scheraga, H. A.: Adv. Phys. Chem. *6*, 103 (1968)
10. Némethy, G.: in: Subunits in Biological Systems, Teil C, Hrsg. G. D. Fasman, S. N. Timasheff, M. Dekker, New York, 1975
11. Scheraga, H. A.: Chem. Rev. *71*, 195 (1971)
12. Hopfinger, A. J.: Conformational Properties of Macromolecules, Academic Press, New York, 1973
13. Brant, D. A.: Ann. Rev. Biophys. Bioenerg. *1*, 369 (1972)
14. Pullmann, B., Pullmann, A.: Adv. Protein Chem. *28*, 347 (1974)
15. Scheraga, H. A.: Biopolymers *20*, 1877 (1981)

16. Scheraga, H. A.: Biopolymers 22, 1 (1983)
17. Blout, E. R., de Lozé, C., Bloom, S. M., Fasman, G. D.: J. Am. Chem. Soc. 82, 3787 (1960)
18. Ramakrishnan, C., Soman, K. V.: Int. J. Peptide Protein Res. 20, 218 (1982)
19. Chou, P. Y., Fasman, G. D.: Adv. Enzymol. 47, 45 (1978)
20. Chou, P. Y., Fasman, G. D.: Biochemistry 13, 212 (1974)
21. Chou, P. Y., Fasman, G. D.: Biochemistry 13, 222 (1974)
22. Chou, P. Y., Fasman, G. D.: J. Mol. Biol. 115, 135 (1977)
23. Chou, P. Y., Fasman, G. D.: Ann. Rev. Biochem. 47, 251 (1978)
24. Kolaskar, A. S., Ramabrahman, V.: Int. J. Peptide Protein Res. 22, 83 (1983)
25. Lifson, S., Sander, C.: Nature 282, 109 (1979)
26. Geisow, M. J., Roberts, R. D. B.: Int. J. Biol. Macromol. 2, 387 (1980)
27. Palau, J., Argos, P., Puigdomenech, P.: Int. J. Peptide Protein Res. 19, 349 (1982)
    Kabsch, W., Sander, Ch.: FEBS Letters 155, 179 (1983)
28. Tanaka, S., Scheraga, H. A.: Macromolecules 9, 168 (1976)
29. Tanaka, S., Scheraga, H. A.: Macromolecules 9, 142 (1976)
30. Tanaka, S., Scheraga, H. A.: Macromolecules 9, 159 (1976)
31. Tanaka, S., Scheraga, H. A.: Macromolecules 9, 812 (1976)
32. Tanaka, S., Scheraga, H. A.: Macromolecules 10, 9 (1977)
33. Tanaka, S., Scheraga, H. A.: Macromolecules 10, 305 (1977)
34. Schiffer, M., Edmundson, A. B.: Biophys. J. 7, 121 (1967)
35. Schiffer, M., Edmundson, A. B.: Biophys. J. 8, 29 (1968)
36. Lim, V. I.: J. Mol. Biol. 88, 857 (1974)
37. Lim, V. I.: J. Mol. Biol. 88, 873 (1974)
38. Schulz, G. E.: Nature 280, 140 (1974)
39. Matthews, B. W.: Biochim. Biophys. Acta 405, 442 (1975)
40. Wolfenden, R. V., Cullis, P. M.: C.C.F. Southgate, Science 206, 575 (1979)
41. Argos, P., Palau, J.: Int. J. Peptide Protein Res. 19, 380 (1982)
42. Sternberg, M. J. E., Thornton, J. M.: Nature 271, 15 (1978)
43. Fischer, E., Otto, E.: Ber. dtsch. Chem. Ges. 36, 2106 (1903)
44. Fischer, E.: Ber. dtsch. Chem. Ges. 39, 530 (1906)
45. Meienhofer, J.: Biopolymers 20, 1761 (1981)
46. Bodanszky, M., Martinez, J.: Synthesis 1981, 333
47. Kovacs, J., Holleran, E. M., Hui, K. Y.: J. Org. Chem. 45, 1060 (1980)
48. Bodanszky, M., Ondetti, M. A.: Peptide Synthesis, Interscience Publ., New York, 1966
49. Schröder, E., Lübke, K.: The Peptides, Vol. 1, Academic Press, New York—London, 1965
50. Merrifield, R. B.: J. Am. Chem. Soc. 85, 2149 (1963)
51. Birr, C., in: Reactivity and Structure, Concepts in Organic Chemistry, Vol. 8, Springer-Verlag, Heidelberg, 1978
52. Barany, G., Merrifield, R. B., in: The Peptides, Vol. 2, Hrsg. E. Gross und J. Meienhofer, Academic Press, New York, 1980, S. 1
53. Bayer, E., Eckstein, H., Hägele, K., König, W. A., Brüning, W., Hagenmaier, H., Parr, W.: J. Am. Chem. Soc. 92, 1735 (1970)
54. Bayer, E., Hagenmaier, H., Jung, G., Parr, W., Eckstein, H., Hunziker, P., Sievers, R. E., in: Peptides 1969, North-Holland Publishing Company, Amsterdam, 1971, S. 65
55. Bayer, E., Hagenmaier, H., Jung, G., König, W., in: Peptides 1968, North-Holland Publishing Company, Amsterdam, 1968, S. 162
56. Atherton, E., Gait, M. J., Sheppard, R. C., Williams, B. J.: Bioorganic Chem. 8, 351–370 (1979)
57. Becker, H., Lucas, H. W., Maul, J., Pillai, V. N. R., Anzinger, H., Mutter, M.: Makromol. Chem., Rapid Commun. 3, 217 (1982)
58. Tam, J. P., in: Peptides, Synthesis-Structure-Function, Hrsg. D. H. Rich, E. Gross, Pierce Chemical Company, Rockford, Illinois, 1981, S. 153
59. Tjoeng, F. S., Heavner, G. A.: Synthesis 1981, 897
60. Tam, J. P., Tjoeng, F. S., Merrifield, R. B.: J. Am. Chem. Soc. 102, 6117 (1980)
61. Mutter, M., Hagenmaier, H., Bayer, E.: Angew. Chem. Int. Ed. Engl. 10, 811 (1971)

62. Bayer, E., Mutter, M.: Nature (London) *273*, 512 (1972)
63. Mutter, M., Bayer, E.: Angew. Chem. *86*, 101 (1974)
64. Mutter, M., Uhmann, R., Bayer, E.: Liebig's Ann. Chem. *1975*, 901
65. Mutter, M., Bayer, E., in: The Peptides, Analysis, Synthesis and Biology, Vol. II, Hrsg. E. Gross, J. Meienhofer, Academic Press, New York, 1980, S. 285
66. Mutter, M.: Tetrahedron Letters *19*, 2839 (1978)
67. Pillai, V. N. R.: Synthesis *1980*, 1
68. Colombo, R.: Tetrahedron Letters *22*, 4129 (1981)
69. Colombo, R., Pinelli, A.: Hoppe-Seylers's Z. Physiol. Chem. *362*, 1385 (1981)
70. Tjoeng, F. S., Heavner, G. A.: Tetrahedron Lett. *23*, 4439 (1982)
71. Tjoeng, F. S., Heavner, G. A.: J. Org. Chem. *48*, 355 (1983)
72. Pillai, V. N. R., Mutter, M., in: Synthetic and Structural Problems, (Topics in Current Chemistry, Vol. 106), Springer-Verlag, Heidelberg, 1982, S. 119
73. Mutter, M.: Habilitationsschrift, Universität Tübingen, 1976
74. Bayer, E., Mutter, M.: Chem. Ber. *107*, 1344 (1974)
75. Mutter, M., Bayer, E.: Angew. Chem. Int. Ed. Engl. *13*, 88 (1974)
76. Pillai, V. N. R., Mutter, M.: Acc. Chem. Res. *14*, 122 (1981)
77. Pillai, V. N. R., Mutter, M.: Naturwissenschaften *68*, 558 (1981)
78. Toniolo, C., Bonora, G. M., Mutter, M.: J. Am. Chem. Soc. *101*, 450 (1979)
79. Mutter, M., Mutter, H., Uhmann, R., Bayer, E.: Biopolymers *15*, 917 (1976)
80. Goodman, M., Saltman, R. P.: Biopolymers *20*, 1929 (1981)
81. Schoknecht, W., Albert, K., Jung, G., Bayer, E.: Liebig's Ann. Chem. *1982*, 1514
82. Leibfritz, D., Mayr, W., Oekonomopulos, R., Jung, G.: TetrahMdron *34*, 2045 (1978)
83. Ribeiro, A. A., Saltman, R. P., Goodman, M., Mutter, M.: Biopolymers *21*, 2225 (1982)
84. Mutter, M.: Macromolecules *10*, 1413 (1977)
85. Mutter, M., Mutter, H., Bayer, E., in: Peptides, Proc. 5th Amer. Pept. Symp. Hrsg. M. Goodman, J. Meienhofer, John Wiley & Sons, New York, 1977, S. 403
86. Goodman, M., Toniolo, C., Naider, F., in: Peptides, Polypeptides and Proteins, Hrsg. E. R. Blout, F. A. Bovey, M. Goodman, N. Lotan, John Wiley & Sons, New York, 1974, S. 308
87. Ingwall, R. T., Goodman, M., in: International Review of Science, Amino Acids, Peptides and Related Compounds, Hrsg. H. N. Rydon, Org. Chem. Ser. 2, Bd. 6, Butterworth, London, 1976, S. 153
88. Naider, F., Goodman, M.: Bioorg. Chem. *3*, 177 (1977)
    C. Toniolo, CRC Crit. Rev. Biochem. *9*, 1 (1980)
89. Toniolo, C., Bonora, G. M., Salardi, S., Mutter, M.: Macromolecules *12*, 620 (1979)
90. Rinaudo, M., Domard, A.: J. Am. Chem. Soc. *98*, 6360 (1976)
91. Toniolo, C., Bonora, G. M., Mutter, M., Pillai, V. N. R.: Makromol. Chem. *182*, 2007 (1981)
92. Toniolo, C., Bonora, G. M., Anzinger, H., Mutter, M.: Macromolecules *16*, 147 (1983)
93. Toniolo, C., Bonora, G. M.: Makromol. Chem. *175*, 2203 (1974)
94. Toniolo, C., Bonora, G. M.: Makromol. Chem. *176*, 2547 (1975)
95. Bonora, G. M., Toniolo, C., Mutter, M.: Polymer *19*, 1382 (1978)
96. Howard, J. C., Cardinaux, F., Scheraga, H. A.: Biopolymers *16*, 2029 (1977)
97. Toniolo, C., Bonora, G. M., in: Peptides: Chemistry, Structure and Biology, Hrsg. R. Walter, J. Meienhofer, Ann Arbor Science, Ann Arbor, 1975, S. 145
98. Miyazawa, T., in: Poly-α-amino Acids, G. D. Fasman, Ed., Marcel Dekker, New York, 1967, p. 69
99. Ciardelli, F., Pieroni, O.: Chimia *34*, 301 (1980)
100. Ciardelli, F., Chiellini, E., Carlini, C., Pieroni, O., Salvador, P., Menicagli, R.: J. Polymer Sci.: Polymer Symposium *62*, 143 (1978)
101. Woody, R. W.: Biopolymers *17*, 1451 (1978)
102. Peggion, E., Cosani, A., Terbojevich, M., Palumbo, M., in: Optically Active Polymers, Hrsg. E. Seleguy, D. Reidel Publishing Company, 1979, S. 231
103. Toniolo, C., Bonora, G. M., Anzinger, H., Mutter, M.: J. Chem. Soc. Chem. Comm., im Druck

104. Toniolo, C., Bonora, G. M., Mutter, M., Pillai, V. N. R.: Makromol. Chem. *182*, 1997 (1981)
105. Toniolo, C., Bonora, G. M., Mutter, M.: Int. J. Biolog. Macromolecules *1*, 188 (1979)
106. Balcerski, J. S., Pysh, E. S., Bonora, G. M., Toniolo, C.: J. Am. Chem. Soc. *98*, 3470 (1976)
107. Toniolo, C., Bonora, G. M., Palumbo, M., Peggion, E., Stevens, E. S.: Biopolymers *17*, 1713 (1978)
108. Liang, J. N., Stevens, E. S., Toniolo, C., Bonora, G. M., in: Peptides 1979, Hrsg. Ed. Gross, J. Meienhofer, Pierce Chemical Company Publishers, Rockford, Illinois, 1980, S. 245
109. Toniolo, C., Bonora, G. M., Crisma, M.: Makromol. Chem. *182*, 3149 (1981)
110. Kelly, M. M., Pysh, E. S., Bonora, G. M., Toniolo, C.: J. Am. Chem. Soc. *99*, 3264 (1977)
111. Toniolo, C., Palumbo, M.: Biopolymers *16*, 219 (1977)
112. Baron, M. H., de Lozé, C., Toniolo, C., Fasman, G. D.: Biopolymers *18*, 411 (1979)
113. Toniolo, C., Bonora, G. M., Palumbo, M., Pysh, E. S., in: Peptides 1976, Hrsg. A. Loffet, Editions de l'Université de Bruxelles, 1976, S. 597
114. Palumbo, M., Da Rin, S., Bonora, G. M., Toniolo, C.: Makromol. Chem. *177*, 1477 (1976)
115. Bonora, G. M., Moretto, V., Toniolo, C., Anzinger, H., Mutter, M.: Int. J. Peptide Protein Res. *21*, 336 (1983)
116a. Kessler, H.: Angew. Chem. *94*, 509 (1982)
116b. Kricheldorf, H. R., Müller, D.: Int. J. Biol. Macromol. *6*, 145 (1984)
117. Ovchinnikov, Y., Ivanov, V. T.: Tetrahedron *30*, 1871 (1974)
118. Ovchinnikov, Y., Ivanov, V. T.: Tetrahedron *31*, 2177 (1975)
119. Bara, Y. A., Friedrich, A., Kessler, H., Molter, M.: Chem. Ber. *111*, 1045 (1978)
120. Kopple, K. D., Ohnishi, M., Go, A.: J. Am. Chem. Soc. *91*, 4264 (1969)
121. Kessler, H., Kondor, P.: Chem. Ber. *112*, 3541 (1979)
     L. G. Pease, C. H. Nin, G. Zimmermann, J. Am. Chem. Soc. *101*, 184 (1979)
122. Kessler, H., Hölzemann, G.: Liebig's Ann. Chem. *1981*, 2028
123. Stevens, E. S., Sugawara, N., Bonora, G. M., Toniolo, C.: J. Am. Chem. Soc. *102*, 7048 (1980)
124. Naider, F., Ribeiro, A. A., Goodman, M.: Biopolymers *19*, 1791 (1980)
125. Goodman, M., Ueyama, N., Naider, F., Gilon, C.: Biopolymers *14*, 915 (1975)
126. Ribeiro, A., Saltman, R. P., Goodman, M.: Biopolymers *19*, 1771 (1980)
127. Asakura, T.: Makromol. Chem. *182*, 1153 (1981)
128. Pitner, T. P., Urry, D. W.: Biochemistry *11*, 4132 (1972)
129. Asakura, T.: Makromol. Chem. *182*, 1135 (1981)
130. Bystrov, V. F., Portnova, S. L., Tsetlin, V. I., Ivanov, V. T., Ovchinnikov, Y. A.: Tetrahedron *25*, 493 (1969)
131. Kopple, K. D., Go, A.: J. Am. Chem. Soc. *99*, 7698 (1977)
132. Kopple, K. D.: Biopolymers *20*, 1913 (1981)
133. Kopple, K. D., Go, A., Pilipauskas, D. R.: J. Am. Chem. Soc. *97*, 6830 (1975)
134. Glickson, J. D., Dadok, J., Marshall, G. R.: Biochemistry *13*, 11 (1974)
135. Pitner, T. P., Glickson, J. D., Dadok, J., Marshall, G. R.: Nature *250*, 582 (1974)
136. Waelder, S., Lee, L., Redfield, A. G.: J. Am. Chem. Soc. *97*, 2927 (1975)
137. Forsén, S., Hoffman, R. A.: J. Chem. Phys. *39*, 2892 (1963)
138. von Dreele, P. H., Brewster, A. I., Dadok, J., Scheraga, H. A., Bovey, F. A., Ferger, M. F., Du Vignaud, V.: Proc. Natl. Acad. Sci. USA *69*, 2169 (1972)
139. Bleich, H. E., Glasel, J. A.: J. Am. Chem. Soc. *97*, 6585 (1975)
140. Molday, R. S., Englander, S. W., Kallen, R. G.: Biochemistry *11*, 150 (1972)
141. Walter, R., Smith, C. W., Sarathy, K. P., Pillai, R. P., Krishna, N. R., Lenkinski, R. E., Glickson, J. D., Hruby, V. J.: Int. J. Pept. Protein Res. *17*, 56 (1981)
142. Naider, F., Ribeiro, A. A., Goodman, M.: Biopolymers *19*, 1791 (1980)
143. Goodman, M., Ribeiro, A. A., Naider, F.: Proc. Natl. Acad. Sci. USA *75*, 4647 (1978)
144. Goodman, M., Ueyama, N., Naider, F.: Biopolymers *14*, 901 (1975)
145. Redfield, A. G., Kunz, S. D., Ralph, E. K.: J. Magn. Reson. *19*, 114 (1975)
146. Dadok, J., Sprecher, R. F.: J. Magn. Reson. *13*, 243 (1974)
147. Cutnell, J. D., Dallas, J., Matson, G., La Mar, G. N.: J. Magn. Reson. *41*, 213 (1980)

148. Ribeiro, A., Goodman, M., Naider, F.: Int. J. Peptide Protein Res. *14*, 414 (1979)
149. Ribeiro, A., Goodman, M., Naider, F.: J. Am. Chem. Soc. *100*, 3903 (1978)
150. Naider, F., Sipzner, R., Steinfeld, A. S., Becker, J. M., Ribeiro, A., Goodman, M.: Proc. 6th Am. Peptide Symp., Hrsg. E. Gross, J. Meienhofer, Pierce Chemical Company, Rockford, Ill., S. 185 (1979)
151. Cram, D. J., Trueblood, K. N., in: Host Guest Complex Chemistry I, (Topics in Current Chemistry *98*), Springer Verlag, Heidelberg, 1981, S. 43
152. von Dreele, P. H., Poland, D., Scheraga, H. A.: Macromolecules *4*, 396 (1971)
153. von Dreele, P. H., Lotan, N., Anatharayanan, V. S., Andreatta, R. H., Poland, D., Scheraga, H. A.: Macromolecules *4*, 408 (1971)
154. Anatharayanan, V. S., Andreatta, R. H., Poland, D., Scheraga, H. A.: Macromolecules *4*, 417 (1971)
155. Kidera, A., Mochiznki, M., Hasegawa, R., Hayashi, T., Sato, H., Nakajima, A., Frederickson, R. A., Powers, S. P., Lee, S., Scheraga, H. A.: Macromolecules *16*, 162 (1983)
156. Mutter, M., Anzinger, H., Bode, K., Maser, F., Pillai, V. N. R., in: Chemistry of Peptides and Proteins, Vol. 1, Hrsg. W. Voelter, E. Wünsch, J. Ovchinnikov, V. Ivanov, Walter de Gruyter & Co., New York, 1982, S. 217
157. = loc. cit. 104
158. Abd el Rahman, S., Anzinger, H., Mutter, M.: Biopolymers *19*, 173 (1980)
159. Stewart, W. E., Mandelkern, L., Glick, R. E.: Biochem. *6*, 143 (1967)
160. Ferreti, J. A., Paolillo, L.: Biopolymers *7*, 155 (1969)
161. Quadrifoglio, F., Urry, D. W.: J. Am. Chem. Soc. *90*, 2755 (1968)
162. Ingwell, R. T., Scheraga, H. A., Lotan, N., Berger, A., Katchalski, E.: Biopolymers *6*, 331 (1968)
163. Bonora, G. M., Palumbo, M., Toniolo, C., Mutter, M.: Makromol. Chem. *180*, 1293 (1979)
164. Gibson, K. D., Scheraga, H. A.: Proc. Natl. Acad. Sci. USA *63*, 9 (1969)
165. Gibson, K. D., Scheraga, H. A.: Proc. Natl. Acad. Sci. USA *63*, 242 (1969)
166. Maser, F., Mutter, M., Toniolo, C., Bonora, G. M.: Biopolymers, in press
167. Maser, F., Altmann, K.-H., Mutter, M., Toniolo, C., Bonora, G. M.: Biopolymers, in press
168. Venkatachalam, C. M.: Biopolymers *6*, 1425 (1968)
169. Lewis, P. N., Momany, F. A., Scheraga, H. A.: Biochim. Biophys. Acta *303*, 211 (1973)
170. Maxer, R., Lancelot, G., Spach, G., in: Peptides 1980 (Hrsg. K. Brunfeldt), Proc. 16th Europ. Pept. Symp., Helsingør, 1980, Scriptor, Kopenhagen, 1981, S. 678
171. Reutimann, H., Straub, B., Luisi, P. L., Holmgren, A.: J. Biol. Chem. *256*, 6796 (1981)
172. Holmgren, A.: Trends Biochem. Sci. *6*, 26 (1981)
173. Brahmachari, S. K., Rapaka, R. S., Bhatnagar, R. S., Ananthanargayanan, V. S.: Biopolymers *21*, 1107 (1982)
174. Bode, K., Goodman, M., Mutter, M.: Biopolymers, in press
175. Kolaskar, A. S., Ramabrahman, V., Soman, V. K.: Int. J. Peptide Protein Res. *16*, 1 (1980)
176. Kopple, K. D., Go, A.: Biopolymers *15*, 1701 (1976)
177. Toma, F., Lam-Tanh, H., Piriou, F., Heindl, M. C., Lintner, K., Fermandjian, S.: Biopolymers *19*, 781 (1980)
178. Fasman, G. D., Chou, P. Y., in: Poly-α-amino Acids, G. D. Fasman, Ed. Marcel Dekker, New York, 1967, p. 114
179a. Jaenicke, R.: Angew. Chem. *96*, 385 (1984)
179b. Wetlaufer, D. B.: Proc. Natl. Acad. Sci. USA *70*, 697 (1973)
180. Honig, B., Ray, A., Levinthal, C.: Proc. Natl. Acad. Sci. USA *73*, 1974 (1976)
181. Finkelstein, A. V., Ptitsyn, O. B.: J. Mol. Biol. *103*, 15 (1976)
182. Rose, G. D., Winters, R. H., Wetlaufer, D. B.: FEBS Letters *63*, 10 (1976)
183. Karplus, M., Weaver, D. L.: Nature *260*, 404 (1976)
184. Ptitsyn, O. B., Finkelstein, A. V., in: Protein Folding, Hrsg. E. Jaenicke, Elsevier/North-Holland Biomedical Press 1980, S. 101
185. Dunfield, L. G., Scheraga, H. A.: Macromolecules *13*, 1415 (1980)
186. Némethy, G., Scheraga, H. A.: Quart Rev. Biophys. *10*, 239 (1977)
187. Scheraga, H. A.: Pure Appl. Chem. *36*, 1 (1973)
188. Levinthal, C.: J. Chem. Phys. *65*, 44 (1968)
189. Levinthal, C.: Scient. Amer. *214*, 42 (1966)

190. Karplus, M., Weaver, D. L.: Biopolymers *18*, 1421 (1979)
191. Weaver, D. L.: Biopolymers *21*, 1275 (1982)
192. Phillips, D.: Scient. Amer. *214*, 78 (1966)
193. Rose, G. D.: J. Mol. Biol. *134*, 447 (1979)
194. Yuschok, T. J., Rose, G. D.: Int. J. Peptide Protein Res. *21*, 479 (1983)
195. Mutter, M., Pillai, V. N. R., Anzinger, H., Bayer, F., Toniolo, C., in: Peptides 1980, Hrsg. K. Brunfeld, Scriptor, Kopenhagen, 1981, S. 660
196. Maser, F., Thesis, Ph. D.: Mainz, 1983
197. Miyazawa, T., Higashijima, T.: Biopolymers *20*, 1949 (1981)
198. Stölzer, D., Duntze, W.: Eur. J. Biochem. *65*, 257 (1976)
199. Tanaka, T., Kitka, H., Murakami, T., Narita, K.: J. Biochem. *82*, 1681 (1977)
200. Gund, P., Andose, J. D., Rhodes, J. B., Smith, G. M.: Science *208*, 1425 (1980)
201. Marshall, G. R., Bosshard, H. E., Ellis, R. A., in: Computer Representation and Manipulation of Chemical Information, Hrsg. W. T. Wipke, S. R. Heller, R. J. Feldmann, E. Hyde, John Wiley, New York, 1974, p. 203
202. Freidinger, R. M., in: Peptides, Synthesis-Structure-Function, D. H. Rich, E. Gross, Eds., Pierce Chemical Company, Rockford, Illinois, 1981, p. 673
203. Blout, E. R.: Biopolymers *20*, 1901 (1981)
204. Hruby, V. J., Mosberg, H. I., Sawyer, T. K., Knittel, J. J., Rockway, T. W., Ormberg, J., Darman, P., Chan, W. Y., Hadley, M. E.: Biopolymers *22*, 517 (1983)
205. Freidinger, R. M., Veber, D. F., Perlow, D. S., Brooks, J. R., Saperstein, R.: Science *210*, 656 (1980)
206. Veber, D. F., in: Peptides, Synthesis-Structure-Function, D. H. Rich, E. Gross, Eds., Pierce Chemical Company, Rockford, Illinois, 1981, p. 685
207. Veber, D. F., Holly, F. W., Nutt, R. F., Bergstrand, S., Torchiana, S. J., Glitzer, M. S., Saperstein, R., Hirschmann, R.: Proc. Natl. Acad. Sci. USA *75*, 2636 (1978)

H.-J. Cantow (Editor)
Received April 16, 1984

Author Index Volumes 1–65

Henrici-Olivé, G. and *Olivé, S.:* The Chemistry of Carbon Fiber Formation from Polyacrylonitrile. Vol. 51, pp. 1–60.

Hermans, Jr., J., Lohr, D. and *Ferro, D.:* Treatment of the Folding and Unfolding of Protein Molecules in Solution According to a Lattic Model. Vol. 9, pp. 229–283.

Higashimura, T. and *Sawamoto, M.:* Living Polymerization and Selective Dimerization: Two Extremes of the Polymer Synthesis by Cationic Polymerization. Vol. 62, pp. 49–94.

Hoffman, A. S.: Ionizing Radiation and Gas Plasma (or Glow) Discharge Treatments for Preparation of Novel Polymeric Biomaterials. Vol. 57, pp. 141–157.

Holzmüller, W.: Molecular Mobility, Deformation and Relaxation Processes in Polymers. Vol. 26, pp. 1–62.

Hutchison, J. and *Ledwith, A.:* Photoinitiation of Vinyl Polymerization by Aromatic Carbonyl Compounds. Vol. 14, pp. 49–86.

Iizuka, E.: Properties of Liquid Crystals of Polypeptides: with Stress on the Electromagnetic Orientation. Vol. 20, pp. 79–107.

Ikada, Y.: Characterization of Graft Copolymers. Vol. 29, pp. 47–84.

Ikada, Y.: Blood-Compatible Polymers. Vol. 57, pp. 103–140.

Imanishi, Y.: Synthese, Conformation, and Reactions of Cyclic Peptides. Vol. 20, pp. 1–77.

Inagaki, H.: Polymer Separation and Characterization by Thin-Layer Chromatography. Vol. 24, pp. 189–237.

Inoue, S.: Asymmetric Reactions of Synthetic Polypeptides. Vol. 21, pp. 77–106.

Ise, N.: Polymerizations under an Electric Field. Vol. 6, pp. 347–376.

Ise, N.: The Mean Activity Coefficient of Polyelectrolytes in Aqueous Solutions and Its Related Properties. Vol. 7, pp. 536–593.

Isihara, A.: Intramolecular Statistics of a Flexible Chain Molecule. Vol. 7, pp. 449–476.

Isihara, A.: Irreversible Processes in Solutions of Chain Polymers. Vol. 5, pp. 531–567.

Isihara, A. and *Guth, E.:* Theory of Dilute Macromolecular Solutions. Vol. 5, pp. 233–260.

Iwatsuki, S.: Polymerization of Quinodimethane Compounds. Vol. 58, pp. 93–120.

Janeschitz-Kriegl, H.: Flow Birefrigence of Elastico-Viscous Polymer Systems. Vol. 6, pp. 170–318.

Jenkins, R. and *Porter, R. S.:* Upertubed Dimensions of Stereoregular Polymers. Vol. 36, pp. 1–20.

Jenngins, B. R.: Electro-Optic Methods for Characterizing Macromolecules in Dilute Solution. Vol. 22, pp. 61–81.

Johnston, D. S.: Macrozwitterion Polymerization. Vol. 42, pp. 51–106.

Kamachi, M.: Influence of Solvent on Free Radical Polymerization of Vinyl Compounds. Vol. 38, pp. 55–87.

Kaneko, M. and *Yamada, A.:* Solar Energy Conversion by Functional Polymers. Vol. 55, pp. 1–48.

Kawabata, S. and *Kawai, H.:* Strain Energy Density Functions of Rubber Vulcanizates from Biaxial Extension. Vol. 24, pp. 89–124.

Kennedy, J. P. and *Chou, T.:* Poly(isobutylene-*co*-β-Pinene): A New Sulfur Vulcanizable, Ozone Resistant Elastomer by Cationic Isomerization Copolymerization. Vol. 21, pp. 1–39.

Kennedy, J. P. and *Delvaux, J. M.:* Synthesis, Characterization and Morphology of Poly(butadiene-g-Styrene). Vol. 38, pp. 141–163.

Kennedy, J. P. and *Gillham, J. K.:* Cationic Polymerization of Olefins with Alkylaluminium Initiators. Vol. 10, pp. 1–33.

Kennedy, J. P. and *Johnston, J. E.:* The Cationic Isomerization Polymerization of 3-Methyl-1-butene and 4-Methyl-1-pentene. Vol. 19, pp. 57–95.

Kennedy, J. P. and *Langer, Jr., A. W.:* Recent Advances in Cationic Polymerization. Vol. 3, pp. 508–580.

Kennedy, J. P. and *Otsu, T.:* Polymerization with Isomerization of Monomer Preceding Propagation. Vol. 7, pp. 369–385.

Kennedy, J. P. and *Rengachary, S.:* Correlation Between Cationic Model and Polymerization Reactions of Olefins. Vol. 14, pp. 1–48.

Subject Index